AF700218

Middle School Science

Genetics Workbook

This book is dedicated to my mom and dad, who shared a wonderful combination of inherited and acquired traits.

The clipart, digital papers, and fonts featured in this book are from OldMarket, CarrieStephensArt, Bricks and Border, Catalyst Science, Kelsey Teaches, Clipart Crush, Ron Leishman Digital Toonage, The Magical Gallery, Sarah Pecorino Illustration, and Kimberly Geswein Fonts.

https://www.teacherspayteachers.com/store/oldmarket
https://www.teacherspayteachers.com/store/carriestephensart
https://www.teacherspayteachers.com/store/bricks-and-border
https://www.teacherspayteachers.com/store/catalyst-science
https://www.teacherspayteachers.com/store/kelsey-teaches
https://clipartcrush.com
https://www.teacherspayteachers.com/store/ron-leishman-digital-toonage
https://www.teacherspayteachers.com/store/the-magical-gallery
https://www.teacherspayteachers.com/store/sarah-pecorino-illustration
https://www.teacherspayteachers.com/store/kimberly-geswein-fonts

Table of Contents

Table of Contents

Table of Contents

Table of Contents

Workbook Information

The Goal of This Workbook:

This workbook is open and go, and its goal is to introduce genetics in a fun way in order to develop a solid foundation for students to build upon. Genetics involves a *lot* of complicated vocabulary, so this workbook aims to break down each of the most important terms in a way that will be understood and remembered by students.

Overview of Contents:

The *Middle School Genetics Workbook* provides a suggested pacing guide of at least 40 days of lessons. It contains a variety of materials to learn, practice, and assess various genetics content commonly taught in middle school science. Reading passages, worksheets, activities, a project, experiments, and assessments are included. Topics consist of Gregor Mendel and his experiments, dominant and recessive alleles, genotype and phenotype, Punnett squares, homozygous and heterozygous, inherited and acquired traits, and codominance and incomplete dominance.

A Note for the Learning Partner or Teacher:

Genetics can be a challenging topic for students to understand. A big reason it can be difficult is because the terminology is complex and the vocabulary words build off of each other. That means breakdowns in understanding need to be caught and addressed early to prevent frustration later. This workbook is designed to be used in the order in which it is presented. (Don't skip sections unless you are certain the student understands the material.) Quick assessments are used periodically to check in with students. If an assessment's results do not indicate mastery, it is recommended the student go back and correct any misconceptions before proceeding. Detailed answer keys and a complete glossary of terms are in the back of this workbook to guide you and assist your student.

About the Terminology in This Book:

The term "parent" is used to denote an organism that contributes genetic material to its offspring. There are different family makeups that do not have a genetic link to all parents and those are equally valid and accepted.

Necessary Supplies:

This workbook requires minimal, inexpensive, and easy-to-find materials. Here is the complete supply list for this workbook.

- Scissors
- Glue stick (liquid glue is not advised)
- Crayons or colored pencils (markers are not advised)
- A coin (or a digital coin flip found online)
- A four-sided die (or a random number generator of some sort)
- Craft supplies (optional and entirely dependent on the student's model project)
- PTC paper test strips (typically less than $10 for a pack of 100 and available from online retailers—can be substituted with cilantro for a similar experiment)

Suggested Pacing Guide

A Day-by-Day Guide

This pacing guide can be used to structure your lessons.

- ❑ **Day 1:** Begin exploring genetics with An Introduction to Genetics on pages 11-13.
- ❑ **Day 2:** Learn all about Mendel's pea plants on pages 14-16.
- ❑ **Day 3:** Review Mendel's experiments using The Condensed Version on page 17 and complete Mendel's Flower Color Experiment and Mendel's Pea Plants Summary on pages 18-19.
- ❑ **Day 4:** Read the Dominant & Recessive text on pages 20-21 and complete the Dominant & Recessive Practice on page 22.
- ❑ **Day 5:** Complete Celebrity Symbols of Dominant & Recessive Alleles and Dominant & Recessive Alleles Flower Coloring on pages 23-24.
- ❑ **Day 6:** Challenge yourself with the Dominant & Recessive Card Sorting Activity and strengthen your understanding with the Dominant & Recessive Extra Practice on pages 25-29.
- ❑ **Day 7:** Gather some people and do the Taste Test Experiment on pages 30-32.
- ❑ **Day 8:** Read the Dominant Might Not Equal Common text on page 33 and answer the questions that follow on page 34. Then assess your understanding with the Dominant & Recessive Assessment on page 35.
- ❑ **Day 9:** Read the Genotype & Phenotype text on page 36 and complete the practice worksheet that follows on page 37.
- ❑ **Day 10:** Complete Genotype & Phenotype Worm Coloring on page 38 and Genotypes Plus Adorable Phenotypes on page 39.
- ❑ **Day 11:** Challenge yourself with the Genotype & Phenotype Odd One Out on page 40 and assess your understanding with the Genotype & Phenotype Assessment on page 41.
- ❑ **Day 12:** Read Introducing Punnett Squares and How to Solve Punnett Squares on pages 42-43. Then complete Practice with Punnett Squares on page 44.
- ❑ **Day 13:** Complete Identifying Incorrect Punnett Squares and Predicting Dominant Traits with Punnett Squares on pages 45-46.
- ❑ **Day 14:** Create a fairy and/or a goblin with the activities on pages 47-62.
- ❑ **Day 15:** Read the Homozygous & Heterozygous text on page 63 and complete the Homozygous & Heterozygous Practice and the Corgi Coloring activity on pages 64-65.
- ❑ **Day 16:** Complete the Homozygous & Heterozygous Punnett Squares on pages 66-68.
- ❑ **Day 17:** Make some colorful monsters on pages 69-71.
- ❑ **Day 18:** Begin the Homozygous & Heterozygous Experiment by completing pages 72-73.
- ❑ **Day 19:** Finish the Homozygous & Heterozygous Experiment by completing pages 74-75.
- ❑ **Day 20:** Complete Famous Monster Genetics on pages 76-78 and assess yourself with the Homozygous & Heterozygous Assessment on page 79.

Suggested Pacing Guide

- ❑ **Day 21:** Match genetic terms and definitions with Review of Genetics Concepts on page 80. Then complete Dominant & Recessive Dog Traits on page 81.
- ❑ **Day 22:** Practice definitions with Genetics Vocabulary Review Volume 1 & 2 on pages 82-83.
- ❑ **Day 23:** Complete Farm Animal Genetics on pages 84-86.
- ❑ **Day 24:** Make your own organism with the project on pages 87-91. This is a longer project, so it may be completed over the course of two days.
- ❑ **Day 25:** Read Beyond Just Dominant and Recessive and answer the questions that go with it on pages 92-93.
- ❑ **Day 26:** Complete Getting Your Traits, read Inherited & Acquired Traits, and answer the practice questions on pages 94-96.
- ❑ **Day 27:** Complete Inherited & Acquired Traits Animal Coloring and do the Dog Trait Card Sorting Activity on pages 97-100.
- ❑ **Day 28:** Do My Own Inherited Traits on pages 101-102.
- ❑ **Day 29:** Start the Inherited Traits Survey Activity by asking the questions on page 103 to as many people as you can and carefully recording the answers in the chart. Then graph the information and complete the data analysis on pages 104-105. This is a longer activity, and may need to be completed over the course of two days.
- ❑ **Day 30:** Learn about a trait of your choice using Research a Trait on page 106.
- ❑ **Day 31:** Complete Inherited & Acquired Traits in People and then assess your understanding with the Inherited & Acquired Traits Assessment on pages 107-108.
- ❑ **Day 32:** Learn about Codominance & Incomplete Dominance on pages 109-111.
- ❑ **Day 33:** Use the text you read on Day 32 to answer the questions on page 112. Then color some fancy unicorns on page 113.
- ❑ **Day 34:** Complete Codominance & Incomplete Dominance Cat Coloring on page 114 and then test your understanding with the assessment on page 115.
- ❑ **Day 35:** Play the Genetics Vocabulary Review Card Game on pages 116-130.
- ❑ **Day 36:** Complete the Genetics Vocabulary & Concepts Review on pages 131-132.
- ❑ **Day 37:** Use your critical thinking skills to complete Genetics Odd One Out on page 133. Then use the game cards you cut out on Day 36 to practice the vocabulary again.
- ❑ **Day 38:** Complete Fairy Tale Genetics on pages 134-135.
- ❑ **Day 39:** Do the Genetics Stations Activity found on pages 136-150 to practice all the genetics information you learned one last time before your test.
- ❑ **Day 40:** Test day! Complete the test on pages 151-155. You've got this!
- ❑ **Day 41+** Use the Ideas to Explore Next and the Hands-On Activities & Research Ideas lists on pages 156-157 to continue learning and experiencing genetics.

An Introduction to Genetics

What do you know about genetics and heredity?

- In the space below, use sentences and pictures to record everything you already know about genetics and heredity.

An Introduction to Genetics

Jeepers Creepers

Think about your eye color. Maybe your irises are a gorgeous dark chocolate. Maybe you have peepers so blue they could make a blueberry jealous. Or *maybe* you have one green eye and one hazel eye. Regardless of the exact shade, you inherited your eye color and countless other characteristics from your parents. Heredity is the passing of traits, which are characteristics such as eye color, from parents to offspring. The study of how traits and genes get passed from parents to offspring is called genetics. In short, heredity is parents giving traits to babies; genetics is the study of heredity.

Key Terms

Offspring: They are new organisms produced by a living thing. Offspring are the babies.

Traits: They are the characteristics or features of an organism.

Heredity: It is the passing of traits from parents to offspring.

Genetics: It is the scientific study of heredity.

Giving Peas a Chance

Back in the 1800s there was an Austrian monk named Gregor Mendel. In that time, most biologists believed in a theory of blending inheritance. Basically, they thought offspring were the result of a simple mixing of traits. Imagine red mixing with white and getting pink—that's blending. Based on his own observations, Mendel was not convinced of blending inheritance. He decided to experiment to learn more about how traits were passed down from one generation to the next. Initially, he tried his experiments on honeybees. You can imagine how that worked out for him. Looking to his monastery's garden, he found the perfect subject: pea plants. Peas were quick and easy to grow and had several traits with only two forms. Some of the pea traits were seed color (either yellow or green), seed shape (smooth or wrinkled), height, and flower color (purple or white).

Over the course of nearly a decade, Mendel conducted experiments on tens of thousands of pea plants. Because of the sheer number of plants in his study, he ensured his experiment results were reliable and accurate. Mendel made some surprising discoveries that disproved the theory of blending inheritance and helped him develop the principles of inheritance that form the backbone of our current understanding of genetics.

An Introduction to Genetics

Directions: Answer the following questions in complete sentences.

1. What are offspring? Give examples.

2. What are traits? Give examples.

3. In your own words, what is heredity?

4. In your own words, what is genetics?

5. Who was Gregor Mendel and what did he do?

6. Describe the theory of blending inheritance that Mendel eventually disproved.

7. According to blending inheritance, what kind of offspring would a tall parent and a short parent be expected to have? Explain your answer.

Mendel's Pea Plants

Mendel's Experiments

Gregor Mendel designed a series of similar experiments. He would start with a parental generation he called the "P generation." This parental generation would produce offspring he called the "F_1 generation."* The F_1 generation would in turn have their own offspring called the F_2 generation.

The P generation consisted of purebreds. A purebred is an organism with many generations of the same trait. For example, if a pea plant with purple flowers has parents, grandparents, great grandparents, and great-great grandparents *all* with purple flowers, it would be considered a purebred for that trait. In Mendel's experiments, he crossed two purebreds with *opposite* traits together. He would cross a tall purebred with a short purebred. Or he would cross a wrinkled-seeded pea plant with a smooth-seeded pea plant. The offspring of the P generation, called the F_1 generation, were hybrids. A hybrid is the offspring of two distinctly different parental types. Offspring from the F_1 generation would cross with each other, resulting in the F_2 generation.

Mendel did these types of experiments over and over again, keeping detailed records of the number of offspring with each trait. Within the numbers, he saw patterns that had statistical significance. In other words, Mendel noticed results that could not have happened due to random chance.

General Outline of Mendel's Experiments

P Generation

Purebred parent x Purebred parent with opposite trait

↓

F_1 Generation

pea offspring (hybrids)

↓

F_2 Generation

More pea offspring

Key Terms

Purebred: A purebred is an organism with many generations of the same trait.

Hybrid: A hybrid is the offspring of two distinctly different parental types.

Garden peas *Pisum sativum*

F_1 stands for first filial, which has origins in Latin words meaning sons and daughters.

Mendel's Pea Plants

Pea Height: Tall? Short? Medium? Ginormous?

When Mendel crossed a purebred tall plant with a purebred short plant, the crossing was expected to produce results that adhered to the widely believed theory of blending inheritance. Which is a fancy way of saying a tall plant crossed with a short plant will result in a plant of medium height. Let's examine what *actually* happened in Mendel's height experiment.

Stop and make predictions below.

1. For the P Generation, Mendel crossed a purebred tall with a purebred short. Predict the height of their offspring (the F_1 generation).

2. Then Mendel crossed the resulting F_1 offspring with each other. Predict the height of their offspring (the F_2 generation).

Sketch your predictions below.

x = =

P Generation | F_1 Generation | F_2 Generation

Mendel's Pea Plants

The Results of the Height Experiment

When the purebred tall plant and the purebred short plant were crossed, *all* of their offspring were tall. The short trait seemed to disappear completely in the F_1 generation. However, the short trait *reappeared* in 25% of the offspring of the next generation.

Mendel repeated this experiment again and again, and always got similar results. One trait would appear in all of the offspring of the F_1 generation. While in the F_2 generation, ¾ of the offspring had one trait but ¼ of the offspring displayed the trait that had disappeared in the previous generation.

Different Traits, Same Results

Mendel did not stop with plant height. He found he got similar results no matter which trait he studied. For example, he considered seed shape—either smooth or wrinkled. Same results.

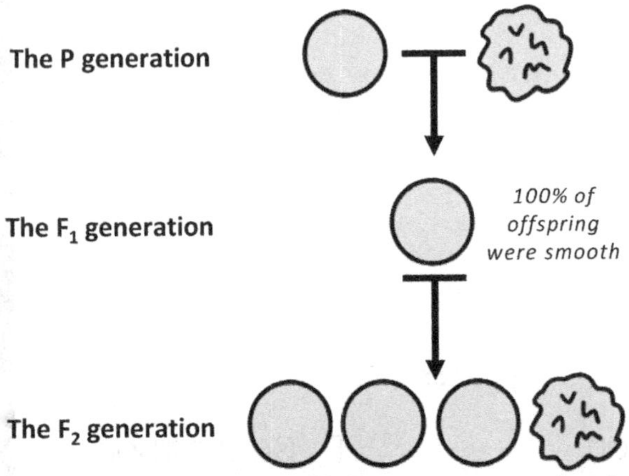

75% of offspring were smooth and 25% of offspring were wrinkled

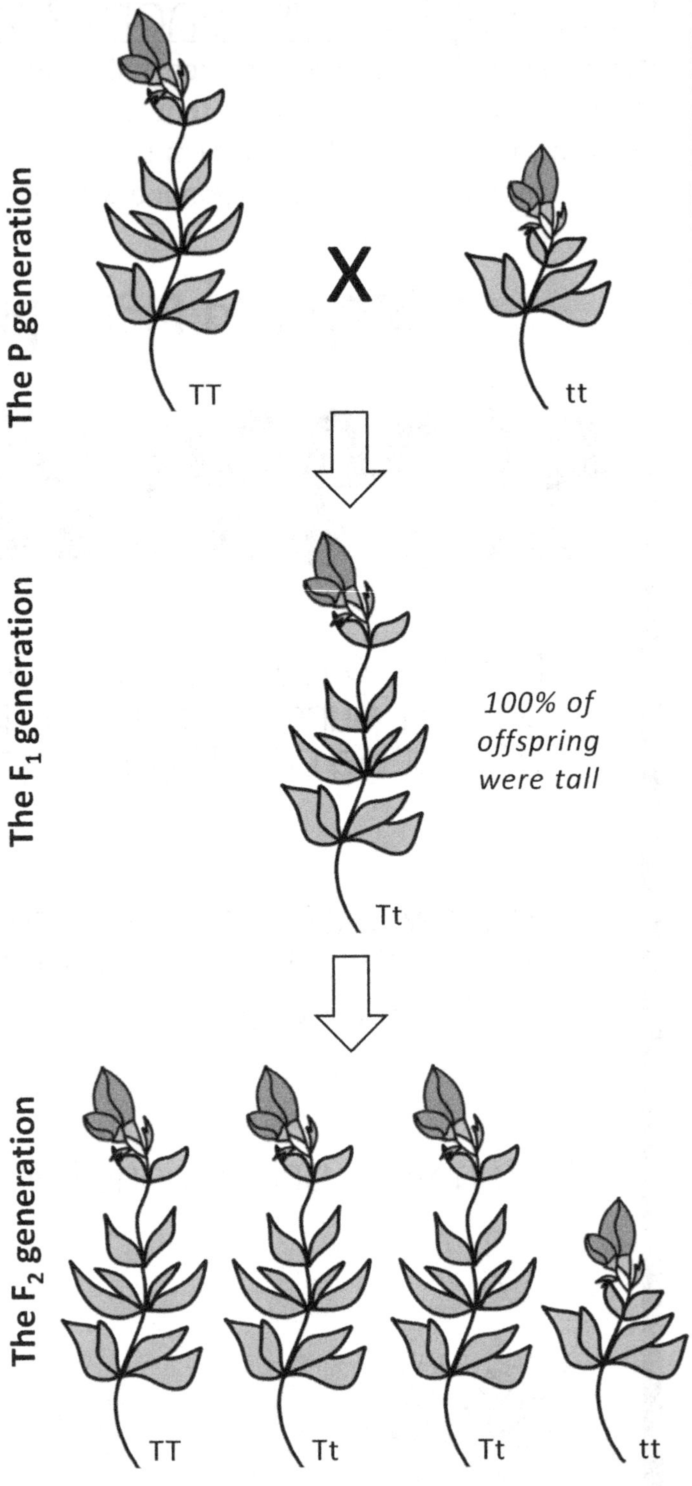

75% of offspring were tall and 25% of offspring were short

Mendel's Pea Plants: The Condensed Version

This tall pea plant came from a family of only tall pea plants.

X

These two plants had babies together.

This short pea plant came from a family of only short pea plants.

All of their babies were tall. The short trait seemed to disappear.

This set of pea plants had babies together.

75% of the offspring were tall and 25% of the offspring were short. Another way to say this is ¾ were tall and ¼ were short.

The short trait came back!

Mendel's Flower Color Experiment

Directions: As you know, Mendel experimented with pea plant height. He also designed a similar experiment with pea plant flower colors where he got the same kind of results. Grab your colored pencils or crayons and apply what you know about Mendel's experiments to complete this page.*

Color this pea plant flower purple. It is a purebred from a family of only purple flowers.

EE

X

These two plants had offspring together.

ee

Leave this pea plant flower white. It is a purebred from a family of only white flowers.

100% of the offspring had ________________ flowers. The ________________ flowers seemed to disappear. Color this flower accordingly.

Ee

This set of pea plants had offspring together.

EE

Ee

Ee

ee

______ % of the offspring were ________________ and ______ % of the offspring were ________________. Color these flowers accordingly.

**If you want to accurately show Mendel's results, you should know that purple flowers are similar to the tall plants and white flowers are similar to the short plants in terms of the displayed traits.*

Mendel's Pea Plants: A Summary

Choose one of Mendel's experiments, either the pea plant height or the pea plant flower color. In your own words, summarize the results of Mendel's experiment. Answer in paragraph form using a minimum of five sentences.

Dominant & Recessive

The Principle of Dominance

One of the big ideas Gregor Mendel developed is called the principle of dominance. This principle states some alleles are dominant and others are recessive; therefore, in a gene, the trait of one allele can hide the trait of the other allele. Hubbawhat?! That probably doesn't make much sense to you. Yet. This principle has some fancy terms that need to be broken *way* down. Get ready from some vocab.

First up are genes. Genes are the basic units of heredity that are passed down from parents to offspring. Genes are in cells, and they contain the information about an organism's traits (characteristics such as hair color and seed shape). Now for the next vocabulary word: Alleles. Alleles are different forms of the genes. If a gene is for height, the allele can be tall or short. If the gene is for flower color, the allele might be purple or white. Each gene contains a pair of alleles, one from each parent. Which brings us to the final two vocabulary words: dominant and recessive.

A dominant allele is one whose trait always shows up if the allele is present. If either parent gives a dominant allele for a trait, then that dominant trait will show up in the offspring. Conversely, a recessive allele is hidden whenever a dominant allele is present. Recessive traits *only* show up if the organism does *not* have a dominant allele for a trait.

Gene

Genes are basic units of heredity passed from parents to offspring. They contain information about traits.

Allele

Alleles are the different versions of a gene. A gene for eye color could have an allele for blue eyes or brown eyes.

Dominant

A dominant allele is one whose trait always shows up if the allele is present.

It DOMINATES!

Recessive

A recessive allele is one whose trait will only show up if there is no dominant allele around.

It hides.

Okay, now let's get back to the principle of dominance and break it down.

- Genes are what give an organism its traits. Genes come in pairs of alleles. One allele is from one parent and the other allele is from the other parent.
- Those alleles can be either dominant (think stronger, beefier, and very much visible), or they can be recessive (think milder, meeker, and happy to hang in the background).
- If there is a dominant allele in a gene, that dominant trait will show up. The only way a recessive trait can be seen is if there is no dominant allele around to make it hide.

Dominant & Recessive

Seeing the Traits

As you now know, a gene has two alleles—one from each parent. The one allele a parent provides might be dominant or recessive. A gene that follows the principle of dominance can have three possible combinations of alleles:

- Two dominant alleles
- One dominant allele and one recessive allele
- Two recessive alleles

The trait that is seen in the offspring is determined by which of the above combinations it receives from its parents. Carefully examine the key points below.

Key Points:

- Two dominant alleles = a dominant trait is seen
- One dominant allele + one recessive allele = a dominant trait is seen
- Two recessive alleles = a recessive trait is seen

Think back to Mendel's pea plant height experiment. When a purebred tall was crossed with a purebred short, all of the offspring were tall. That means tall is the dominant trait for pea plant height, and short is the recessive trait. Mendel needed a quick and easy way to represent that on paper, so he decided to use capital and lowercase letters. **Capital letters are used for the dominant alleles. Lowercase letters are used for the recessive alleles.** The specific letter that is used (think letter *A* or *B* or *C*...) is not all that important. What matters is whether that letter is capitalized (dominant) or lowercased (recessive).*

Flip back to the diagram of Mendel's pea plant height experiment found on page 17. Do you notice the pairs of letter T next to the pictures of pea plants? Those might be making more sense now.

- TT represents two dominant alleles for plant height. Two dominant alleles make a purebred tall plant. Having two dominant alleles is *one way to have a tall* pea plant.
- tt represents two recessive alleles for plant height. Two recessive alleles make a purebred short plant. Having two recessive alleles for height is the *only way to have a short* pea plant.
- Tt represents one dominant and one recessive allele for plant height. One dominant allele with one recessive allele makes a tall pea plant. Having one dominant and one recessive is *another way to have a tall* pea plant.

**If an allele pair has one capital letter and one lowercase letter, the capital letter goes first. It's Bb, not bB.*

Dominant & Recessive Practice

Directions: Answer the following questions in complete sentences.

1. What are genes and where are they located? ______________________________

2. What are alleles? In a pair of alleles, how many alleles come from each parent? __________

3. In your own words, what is the principle of dominance? ______________________________

4. In your own words, what is dominant? ______________________________

5. In your own words, what is recessive? ______________________________

Directions: Read the questions below and circle your answers.

6. If an organism has two dominant alleles (TT) for a trait, which trait will be seen?
   The dominant trait   The recessive trait   A mix of both traits

7. If an organism has two recessive alleles (tt) for a trait, which trait will be seen?
   The dominant trait   The recessive trait   A mix of both traits

8. If an organism has one dominant allele and one recessive allele (Tt) for a trait, which trait will be seen?
   The dominant trait   The recessive trait   A mix of both traits

Celebrity Symbols of Dominant & Recessive Alleles

Quick Review:

1. Are capital letters used to represent dominant or recessive alleles? ____________________
2. Does RR have two dominant alleles or two recessive alleles? ____________________
3. What does Ff represent? ____________________
4. Which trait is seen if both a dominant allele and a recessive allele are present? ____________________
5. What is the only way a recessive trait can be seen? ____________________

Directions: Choose one celebrity to represent dominant and one to represent recessive. There are a lot of different ways you can think about this. You might choose a big, muscular athlete to show dominant. Or maybe your dominant could be a singer with a sunny personality who lights up a room. Draw them in the photos below and write their names. Explain your choices in sentences.

Here's why this is my dominant celebrity:

My recessive celebrity:

My dominant celebrity:

Here's why this is my recessive celebrity:

Dominant & Recessive Alleles Flower Coloring

Directions: Color the flowers describing dominant alleles yellow and the flowers describing recessive alleles red. Then write in complete sentences to answer the question at the bottom of the page.

1

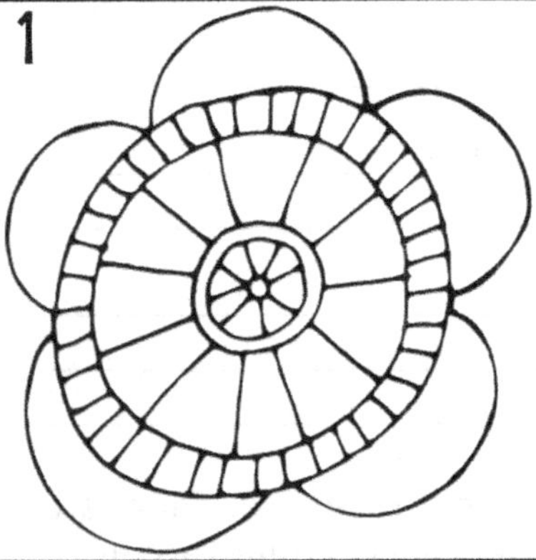

The allele whose trait always shows up if the allele is present

2

The allele that only shows its trait when the other allele is not around

3

They are represented with lowercase letters.

4

They are represented with capital letters.

5

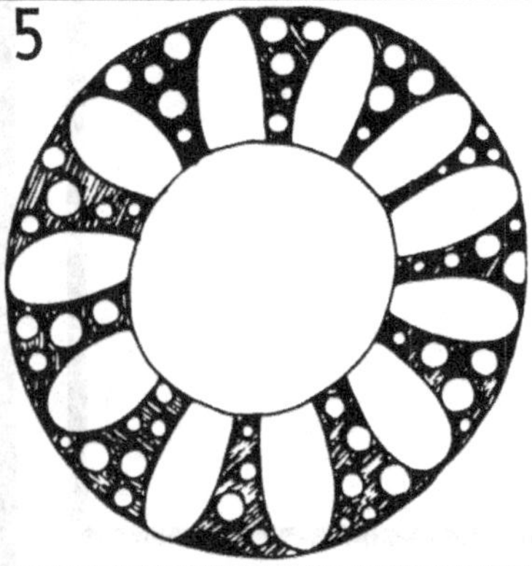

The type of allele whose trait would be visible in this pair: Gg

6

ee

What is the difference between dominant and recessive alleles? ______________________

____

____

____

____

____

____

Dominant & Recessive Card Sorting Activity

Directions:

1. Cut out the cards on page 27.
2. Sort the cards into the chart below.
3. Check your answers to make sure they are correct.
4. Explain your answers to another person. You can use this sentence frame to guide you.
   → "I can tell the _____________________ card goes in the __________________ column of the chart because ________________."

The dominant trait will be seen.	The recessive trait will be seen.

Dominant & Recessive Card Sorting Activity

Directions:

1. Carefully cut out the cards below.
2. Sort the cards into the correct columns of the chart on page 25.
3. Follow the rest of the directions on page 25.

There is one card with a trick answer. Can you find it?

Two dominant alleles paired together	Two recessive alleles paired together
rr	One dominant allele and one recessive allele paired together
Nn	GG
Two capital letters	Two lowercase letters
A capital and a lowercase letter	A hybrid
The offspring has two purebred parents with the same trait.	The offspring has two purebred parents with opposite traits.

Dominant & Recessive Extra Practice

1. Dominant alleles are shown with ______________________ letters.

2. Recessive alleles are shown with ______________________ letters.

3. **There are two ways to show a dominant trait.** Circle the two allele combinations that will result in a dominant trait being seen.

   Two dominant alleles paired together | One dominant allele paired with one recessive allele | Two recessive alleles paired together

4. **There is only one way to show a recessive trait.** Draw a box around the one allele combination that will result in a recessive trait being seen.

   Two dominant alleles paired together | One dominant allele paired with one recessive allele | Two recessive alleles paired together

5. Examine the allele pairs below. Circle all of the allele pairs that will result in a dominant trait being seen.

   Uu   Ll   AA   BB   dd   ee

6. Examine the allele pairs below. Draw a box around all of the allele pairs that will result in a recessive trait being seen.

   Tt   rr   FF   ii   Kk   nn

7. Consider the allele pair Yy, which has one dominant allele (Y) and one recessive allele (y).

   Is there a dominant allele? → *Yes! It's the Y in Yy.*

   Is there a recessive allele? → ____

   Is the dominant trait seen? → *Yes! When there is both a dominant and a recessive allele, only the dominant trait is seen.*

   Is the recessive trait seen? → ____

   Is the recessive allele gone completely **or** is it lurking in the background while the dominant allele gets noticed? → ____

   ____

Taste Test Experiment

Experiment Overview:

In this experiment, people will lick PTC paper test strips to determine if they can taste the chemical phenylthiocarbamide. A person's ability to taste or not taste this chemical is a genetically inherited trait. There are three things that might happen when licking the PTC paper.

- Non-tasters will taste nothing when they lick the paper.
- Tasters will taste a moderately bitter flavor when they lick the paper.
- Supertasters will taste an extremely bitter flavor when they lick the paper.

Experiment Supplies:

- PTC (phenylthiocarbamide) test strips—these are inexpensive and available online from Amazon and other retailers
- As many willing taste-testers as possible—preferably a large, related group of siblings, parents, grandparents, aunts, uncles, and cousins so the trait can be tracked through the family (An unrelated group of people is fine as well.)

Genetic Explanation:

The PTC gene comes in two different alleles, one for tasting and one for non-tasting. Supertasters have two of the same tasting alleles for this trait. Tasters have one tasting allele and one non-tasting allele. Non-tasters have two of the same non-tasting alleles. Notice how the tasters are a mix between the supertasting ability and the inability to taste PTC? This is an example of incomplete dominance—but more on that later in this workbook.

A Note on Safety:

Phenylthiocarbamide is a chemical and in large quantities can be harmful. In order to feel negative effects, a person would need to ingest thousands of these test strips in one go. As the saying goes, the dose makes the poison. To be on the safe side, it is best to keep the pack of these test strips away from small children and pets to prevent accidental ingestion of large amounts of PTC in small bodies. If you are uncomfortable with the risk, a similar experiment can be completed using cilantro and whether or not it tastes soapy.

Taste Test Experiment

Directions: Gather your PTC test strips and your group of willing participants. Provide each person with one test strip and instruct them to lick the paper. Use their reactions and descriptions of the taste to classify them as supertasters, tasters, or non-tasters. Graph the results and note any patterns in the data. If your group of participants is related, make a family tree to trace this trait through the generations.

Name	Supertaster	Taster	Non-Taster
1.			
2.			
3.			
4.			
5.			
6.			
7.			
8.			
9.			
10.			
11.			
12.			
13.			
14.			
15.			
16.			
17.			
18.			
19.			
20.			

Taste Test Experiment

Graph the Results:

1. Write a title above the graph.
2. If you tested more than twenty people, change the interval on the y-axis if needed.
3. Create a bar graph of supertasters, tasters, and non-tasters.
4. Analyze your bar graph and note any trends in the results.

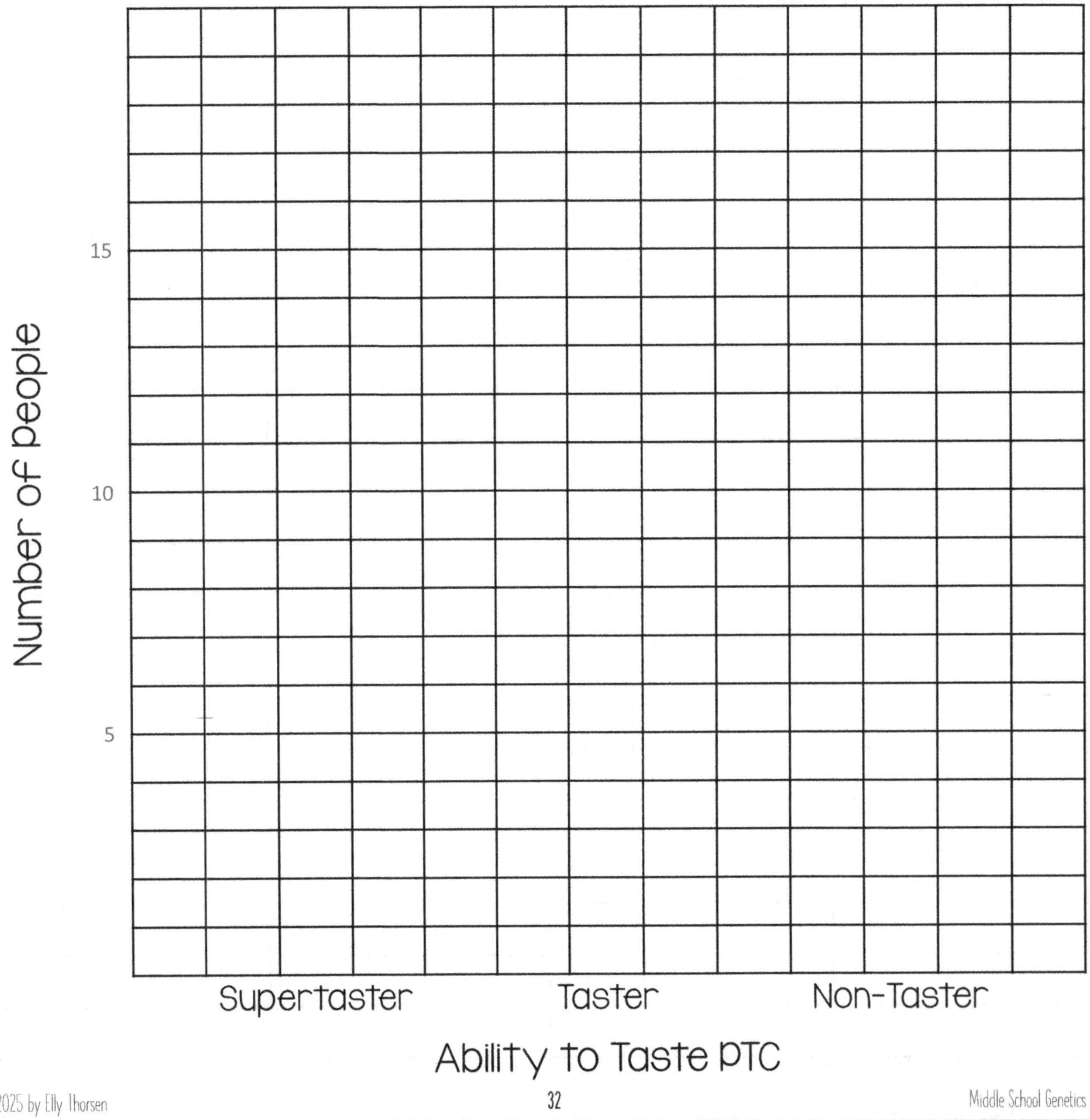

Dominant Might Not Equal Common

A Common Misconception

When dominant alleles are present, the dominant trait will be seen in the offspring. Therefore, it seems logical to assume dominant traits are more common in populations of organisms. This is not always the case. Dominant means it will show up over a recessive; dominant does *not* necessarily mean common.

Regardless of whether a trait is dominant or recessive, the commonness of a trait is due to how many copies of that allele are in the population. If there are very few dominant alleles in a population, then the dominant trait will not be seen very often. The number of copies of an allele in a population might stay fairly stable over time if there is no reason for it to increase or decrease. However, the copies of an allele in a population might increase or decrease due to advantages, disadvantages, random events, or chance.

Natural Selection

If a trait is beneficial for survival or reproduction, then it becomes more common over time. Being incredibly tall might be a recessive trait for an organism, but if height helps it reach more food, then that recessive allele is likely to become more common than the dominant allele. If a trait is harmful for survival or reproduction, then it becomes less common over time. Being born a bright color might be a dominant trait, but if that bright color makes it easier to be seen by predators, then that organism is less likely to reach adulthood and reproduce. The dominant allele for brightly colored babies will likely become less common in that population. The prevalence of traits due to advantages and disadvantages are features of natural selection.

Genetic Drift

If a trait has no impact on survival or reproduction, then it has no reason to become more or less common. It just kind of hangs out in the gene pool and pops up every now and again. The frequency of that trait can increase or decrease due to random chance or events. That's called genetic drift, and it can make a noticeable difference to small populations of organisms. There are two common examples of genetic drift: the bottleneck effect and the founder effect.

The bottleneck effect happens when a population of organisms gets dramatically smaller—maybe due to disease or natural disaster. The surviving organisms might not be representative of the original population, so the genes that circulate in the small new population are different than they were before. Recessive alleles might now be quite common in this population.

The founder effect happens when a small number of individuals leaves a large population and starts up a new population elsewhere. The new colony is made up of a small starting population with limited variation in the gene pool. The recessive alleles might now outnumber the dominant alleles, causing a shift in the population's traits.

Dominant Might Not Equal Common

Directions: First highlight or underline the most important sentence(s) in each of the paragraphs on page 33. Then complete questions 1-6 by answering in sentences.

1. Are dominant traits more common than recessive traits? Use an example and/or details to support your answer. ______________________________

2. Describe one reason why a dominant trait might be less common than a recessive trait.

3. When a trait helps an organism survive or reproduce, what is most likely to happen to that trait in a population over time? Why? ______________________________

4. When a trait makes it more difficult for an organism to survive or reproduce, what is most likely to happen to that trait in a population over time? Why? ______________________________

5. When a trait makes no impact on an organism's ability to survive and reproduce, what is most likely to happen to that trait in an undisturbed population over time? Why? ______________________________

6. What are two examples of genetic drift? Do those examples involve large or small populations of organisms? ______________________________

Dominant & Recessive Assessment

1. True or False: Capital letters are used to represent dominant alleles.

2. What do you call the allele whose trait always shows up in the offspring if it is present?

   ____

3. What do you call the allele that is hidden whenever the other allele is present?

   ____

4. What happens to the recessive allele when a dominant allele is present?
   a. The recessive allele mixes with the dominant allele and both traits are seen.
   b. The recessive allele disappears completely and only the dominant allele's trait is seen.
   c. The recessive allele takes over and covers the dominant allele's trait.
   d. The recessive allele is there but its trait is hidden and only the dominant allele's trait is seen.

5. In a plant with an allele pair of Gg, what trait or traits will show up?
   a. The dominant trait only
   b. The recessive trait only
   c. Both the dominant and recessive traits
   d. Neither the dominant nor recessive traits

6. True or False: An organism with a bb allele pair will show the recessive trait.

7. True or False: Dominant traits are always more common than recessive traits.

Genotype & Phenotype

In Your Genes or in the Mirror

Genotype and phenotype are two words you'll see over and over again when learning about genetics. Like many words related to heredity, they sound trickier than they actually are.

Genotype starts the same way as **gen**e, which should give you a big clue as to its meaning. Genotype is the allele combination or genetic makeup of an organism. Genotype has to do with the allele pairs written using capital and lowercase letters. Some examples of genotypes are WW and Cc and ff. Saying an organism has two dominant alleles for a trait is also an example of genotype. Purebred and hybrid are referring to genotype. Basically, if a combination of alleles is described in any way, you're probably considering a genotype.

Phenotype is an organism's physical appearance or the traits it has—things we can often see. Phenotype is what you are able to observe in an organism. An easy way to remember **ph**enotype is that it starts the same way as **ph**oto, and phenotypes and photographs both show how things appear. Some of your own phenotypes are your hair texture, your skin color, your blood type, and whether you have wet or dry earwax. *Yum.*

More Pea Plants! Yay!

It's time to consider Gregor Mendel and his pea plants again. Let's return to the height experiment. Earlier, you learned tall is dominant and short is recessive. You also learned there are two ways to see the tall dominant trait but only one way to see the short recessive trait.

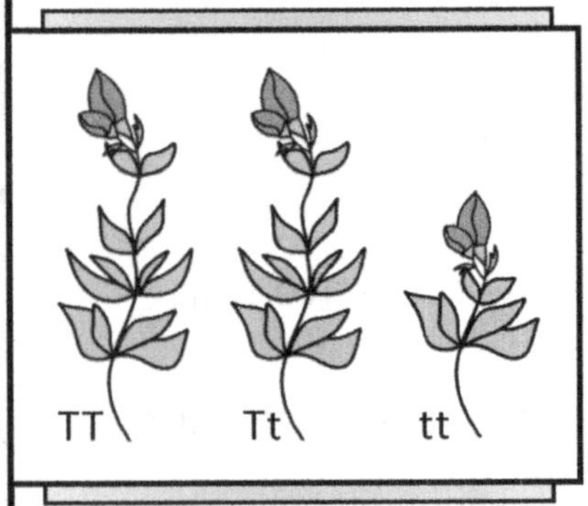

Directions: Use what you have recently learned about genetics to complete the section below about the pea plants pictured on the left.

1. What are the possible genotypes? ______________________________
2. What are the possible phenotypes? ______________________________
3. Tall pea plants can have a genotype of __________ or __________.
4. Short pea plants can only have a genotype of __________.

Directions: For questions five and six, fill in the blanks with the words *genotype* or *phenotype*.

5. All of the tall pea plants have the same ______________________________, but they can have a different ______________________________.
6. All short pea plants have the same ______________________________ and the same ______________________________.

Genotype & Phenotype Practice

Directions: Answer the following in complete sentences.

1. Explain the meaning of genotype in your own words and provide two examples. ________

2. Explain the meaning of phenotype in your own words and provide two examples. ________

Directions: First read the information about the fish. Then read the statements that follow and circle true or false. Finally, explain what makes the statement true or false and/or explain your reasoning.

Fish Information: At your local pet store, you see two fish swimming in an aquarium. You notice both fish have the same dominant triangle pattern on their dorsal fins.

3. **True** or **False**: These fish have the same phenotype. ________

4. **True** or **False**: These fish must have the same genotype. ________

Genotype & Phenotype Worm Coloring

Directions: Color the worms with genotypes green and the worms with phenotypes purple. Then write in complete sentences to answer the question at the bottom of the page.

1

The allele combination or genetic makeup of an organism

2

The physical appearance or the traits of an organism

3

Tall

4

Rr

5

The worm has black hair on its head.

6

Two recessive alleles are paired together.

How can two organisms have the same phenotype but different genotypes? ___________

____

____

____

____

____

____

Genotypes Plus Adorable Phenotypes

Guinea pigs!! Squeeeeeeee! Guinea pigs are pretty stinking adorable. (And have you heard them make their little squeaks and squeals?!) The hair of guinea pigs comes in multiple lengths, colors, and textures, each cuter than the last. For this assignment, though, you'll be focusing on hair length.

Directions: First read the hair length information below. Then draw one guinea pig in each box. The box on the left should show a guinea pig with short hair. The box on the right should show a guinea pig with long hair. After you have drawn your guinea pigs, record their genotypes and phenotypes and answer the question that follows.

Guinea Pig Hair Length Information: In guinea pigs, long hair is recessive and short hair is dominant. In order to have long hair, two recessive alleles (gg) are needed. In order to have short hair, at least one dominant allele needs to be present (Gg or GG).

Genotype(s): ______________________

Phenotype: ______________________

Genotype(s): ______________________

Phenotype: ______________________

What is the difference between genotype and phenotype? Reference your pictures and the information above in your answer. Write in complete sentences. ______________________

____

____

____

____

____

____

____

____

____

Bonus Task: Name your guinea pigs!

Genotype & Phenotype Odd One Out

Directions: Analyze each part of the circle. One of the parts will be different than the others. Shade the part of the circle that does not match. Using a complete sentence, explain how that part is different from the others. *Answers must be science related and should show your understanding of genotype and phenotype.*

Example

Chemistry	History
Biology	Physics

History is the odd one out because it is not a type of science like the others.

1

Genotype	Rr
Allele combination	Visible traits

2

Two recessive alleles	Two dominant alleles
Heart-shaped leaf	hh

3

The genetic makeup of an organism	Physical characteristics
One dominant allele and one recessive allele	Mm

4

Purebred	Pair of recessive alleles
Letters	Tall

5

Hybrid	Bright orange exoskeleton
Wrinkled wings	Short antenna

6

A dog has wiry fur.	A dog has one blue eye and one brown eye.
A dog's ancestors are all the same breed.	A dog's shoulder height is 52 cm.

Genotype & Phenotype Assessment

1. A ______________________________ is the allele combination or genetic makeup of an organism.

2. A ______________________________ is the organism's physical appearance or the trait it has.

3. What is the difference between genotype and phenotype?
   a. Genotype is the genetic makeup of an organism and is represented with letters, while phenotype is the trait of an organism and is often visible.
   b. Phenotype is the genetic makeup of an organism and is represented with letters, while genotype is the trait of an organism and is often visible.
   c. Genotype and phenotype mean the same thing.
   d. Genotype is when an organism has two of the same alleles for a trait, and phenotype is when an organism has two different alleles for a trait.

Directions: Identify each example as a genotype or a phenotype by circling your answer.

4. Genotype or Phenotype: MM

5. Genotype or Phenotype: Brown eyes

6. Genotype or Phenotype: Curly hair

7. Genotype or Phenotype: Oo

8. Genotype or Phenotype: A pair of two recessive alleles

9. Genotype or Phenotype: That is a purebred litter of puppies.

10. Genotype or Phenotype: Every kitten in the litter has striped fur.

Introducing Punnett Squares

Good Ol' Reginald

Two decades after Gregor Mendel died, a British geneticist with a supercool name made a special diagram he named after himself. In 1905, Reginald Crundall Punnett created a "Punnett square" that could predict the genetic makeup of offspring. Using the genotypes of the parents, a Punnett square will show the possibility of their offspring inheriting a certain trait. Read the information in the first gray rectangle to examine how Punnett squares work. One note before you begin: Parents are located on the *outside* of a Punnett square. Offspring are on the *inside* of a Punnett square.

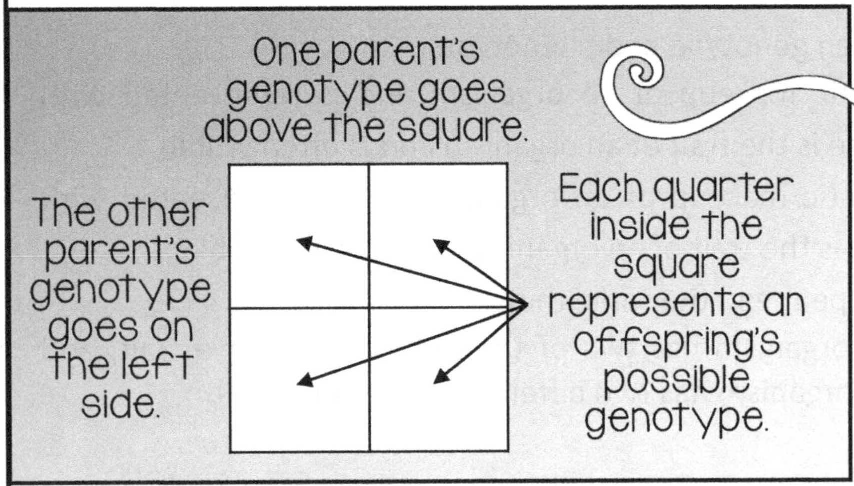

Peas Revisited

It's time to consider pea plants again. Hooray! Remember how in Mendel's height experiment, he began with a tall purebred (TT) and a short purebred (tt)? We can show their offspring in a Punnett square. Behold, the crossing of the P generation!

The P generation's possible offspring are shown in the completed Punnett square. All of the offspring have the same genotype of Tt. Another way to say that is four out of four offspring are expected to have a genotype of Tt. And *another* way to say that is 100% of the offspring are expected to have a genotype of Tt.

To link this up with what you have learned about genetics, you know the dominant trait is tall and the recessive trait is short. When a dominant and a recessive allele are paired together, only the dominant trait is seen. Therefore, 100% of the offspring from the P generation will have a tall phenotype.

How to Solve Punnett Squares

Your Step-by-Step Guide

Follow the written steps to solve a Punnett square. The large box below demonstrates each step.

1. Identify each parent's genotype. *Sometimes this information is provided directly and other times you have to use clues to guide you.*
2. Write one parent's genotype on top of the Punnett square. *It doesn't matter which parent goes on top; the overall results will be the same.*
3. Write the other parent's genotype on the left side of the Punnett square.
4. Start with the top left letter. Write that letter in each of the two boxes directly below it.
5. Move on to the top right letter. Write that letter in each of the two boxes directly below it.
6. Go on to the letters on the left side of the Punnett square. Consider the top letter. Write that letter in each of the two boxes directly to the right. *Remember: capital letters go in front of lowercase letters. Think of it as a grammar rule you must follow.*
7. Now for the final letter, the one on the bottom left of the Punnett square. Write that letter in each of the two boxes directly to the right. The Punnett square is now complete.
8. Use the results of the completed Punnett square to predict the offspring.

Step 1:

The genotypes of the parents are Hh and Hh.

Step 2:

	H	h

Step 3:

	H	h
H		
h		

Step 4:

	H	h
H	H	
h	H	

Step 5:

	H	h
H	H	h
h	H	h

Step 6:

	H	h
H	HH	Hh
h	H	h

Step 7:

	H	h
H	HH	Hh
h	Hh	hh

Step 8:

1 out of 4 have HH.
25% chance of HH

2 out of 4 have Hh.
50% chance of Hh

1 out of 4 have hh.
25% chance of hh

Practice with Punnett Squares

Directions: Solve each of the Punnett squares below. Then use the results to predict the offspring. The first one is done for you. Help calculating percent is at the bottom of this page if needed.

1. **The parents are RR and rr.**

	R	R
r	Rr	Rr
r	Rr	Rr

0 ____ out of 4 are RR
0 ____ % chance of RR

4 ____ out of 4 are Rr
100 ____ % chance of Rr

0 ____ out of 4 are rr
0 ____ % chance of rr

2. **The parents are Aa and aa.**

____ out of 4 are AA
____ % chance of AA

____ out of 4 are Aa
____ % chance of Aa

____ out of 4 are aa
____ % chance of aa

3. **The parents are EE and EE.**

____ out of 4 are EE
____ % chance of EE

____ out of 4 are Ee
____ % chance of Ee

____ out of 4 are ee
____ % chance of ee

4. **The parents are mm and Mm.**

____ out of 4 are MM
____ % chance of MM

____ out of 4 are Mm
____ % chance of Mm

____ out of 4 are mm
____ % chance of mm

5. **The parents are Bb and Bb.**

____ out of 4 are BB
____ % chance of BB

____ out of 4 are Bb
____ % chance of Bb

____ out of 4 are bb
____ % chance of bb

6. **The parents are FF and Ff.**

____ out of 4 are FF
____ % chance of FF

____ out of 4 are Ff
____ % chance of Ff

____ out of 4 are ff
____ % chance of ff

Need help with percent? In order to calculate percent, you divide the part by the whole and then multiply by 100.

4 ÷ 4 = 1	3 ÷ 4 = 0.75	2 ÷ 4 = 0.5	1 ÷ 4 = 0.25	0 ÷ 4 = 0
1 x 100 = 100%	0.75 x 100 = 75%	0.5 x 100 = 50%	0.25 x 100 = 25%	0 x 100 = 0%

Identifying Incorrect Punnett Squares

Directions: Examine each of the Punnett squares below. Some of the Punnett squares are correct and others contain a mistake within the Punnett square.

- If a Punnett square is correct: Circle the word *correct*. You're done. Move on to the next question.
- If any part of a Punnett square contains a mistake: Circle the word *incorrect*, X out the mistake, and write the correct answer.

1.

	B	b
b	Bb	bb
b	BB	bb

Correct

Incorrect

2.

	L	L
l	Ll	Ll
l	Ll	ll

Correct

Incorrect

3.

	E	e
E	EE	Ee
e	Ee	ee

Correct

Incorrect

4.

	r	r
R	Rr	Rr
r	Rr	rr

Correct

Incorrect

5.

	t	t
t	tt	tt
t	tt	tt

Correct

Incorrect

6.

	N	n
N	NN	Nn
n	NN	nn

Correct

Incorrect

Predicting Dominant Traits with Punnett Squares

Directions: Use the provided information to solve a Punnett square. (You can choose what letter of the alphabet to use.) Use the results of the Punnett square to predict the chances of the offspring displaying the dominant phenotype. Write your answer in percent form. The first is done for you.

1. The dominant trait is striped. The recessive trait is spotted. One parent has one dominant and one recessive allele. The other parent has two recessive alleles. Predict the chances of this cross producing the dominant striped phenotype.

   The dominant striped phenotype can show up with either a LL or a Ll genotype. There are no LL genotypes, but there are two Ll genotypes. Two out of four is 50%. ***There is a 50% chance of producing the dominant striped phenotype.***

	L	l
l	Ll	ll
l	Ll	ll

2. The dominant trait is calm. The recessive trait is energetic. One parent has two dominant alleles. The other parent has one dominant and one recessive allele. Predict the chances of this cross producing the dominant calm phenotype.

3. The dominant trait is green. The recessive trait is blue. One parent is a purebred with two dominant alleles. The other parent is a purebred with two recessive alleles. Predict the chances of this cross producing the dominant green phenotype.

4. The dominant trait is smooth. The recessive trait is rough. Both parents are hybrids with one dominant and one recessive allele each. Predict the chances of this cross producing the dominant smooth phenotype.

Make a Fairy Activity

Activity Overview:

You will randomly select the traits of two fairy parents and solve Punnett squares to determine your fairy's inherited traits. Then you will color, cut, and glue your fairy together onto a woodland scene you have colored.

Needed Supplies:

- Scissors
- Glue stick
- Colored pencils or crayons
- A coin or a digital coin flip
- A four-sided die or other method of randomly generating a number between one and four

Activity Directions:

1. Follow the directions on page 48 to determine the parents' genotypes and phenotypes.
2. Solve the Punnett squares on page 49 and record your fairy's genotypes and phenotypes.
3. Color the woodland scene on page 53.
4. Color your fairy and all of your fairy's inherited phenotypes on page 51 according to Trait 1's phenotype for coloration (either pastel or bright colors).
5. Carefully cut out the fairy and all of its phenotypes.
6. Glue your fairy onto the woodland scene in this order: Wings first, body second, face third, hair fourth, and shoes last.

Make a Fairy Activity:

Parents' Genotypes & Phenotypes

Directions: Discover your fairy's parents' genotypes with coin flips. Heads is dominant and tails is recessive. To find each parent's genotype, flip a coin two times to determine what pair of alleles they have for a trait. Use the genotype to determine that parent's phenotype and record the genetic information in the table below. *Example: If your first coin flip is heads (dominant) and your second coin flip is tails (recessive), the genotype of the first parent would be Tt. For Trait 1, that parent's phenotype would be pastel colors.*

Trait 1: Coloration

- Pastel colors are dominant.
- Bright, vibrant colors are recessive.

Trait 2: Wings

- Feathered wings are dominant. 
- Ornate butterfly wings are recessive. 

Trait 3: Personality

- Kind personalities are dominant. 
- Evil personalities are recessive. 

Trait 4: Hair

- Straight hair is dominant. 
- Curly hair is recessive. 

Trait 5: Toe Shape

- Toes that curl up are dominant. 
- Straight toes are recessive.

		Parent 1	Parent 2
Trait 1	Genotype		
	Phenotype		
Trait 2	Genotype		
	Phenotype		
Trait 3	Genotype		
	Phenotype		
Trait 4	Genotype		
	Phenotype		
Trait 5	Genotype		
	Phenotype		

Make a Fairy Activity:

Your Fairy's Genotypes & Phenotypes

Directions: Solve the Punnett squares for each of the five traits using the parents' genotypes you recorded on page 48. After you have solved the Punnett squares, double-check to make sure you solved correctly. Then roll a four-sided die to randomly select which quarter of the Punnett square contains your fairy's genotype. Make a new roll for each of the five traits. Use the trait information on page 48 to determine your fairy's phenotypes. Record your fairy's genetic information in the chart.

Trait 1

1	2
3	4

Trait 2

1	2
3	4

Trait 3

1	2
3	4

Trait 4

1	2
3	4

Trait 5

1	2
3	4

		Your Fairy
Trait 1	Genotype	
	Phenotype	
Trait 2	Genotype	
	Phenotype	
Trait 3	Genotype	
	Phenotype	
Trait 4	Genotype	
	Phenotype	
Trait 5	Genotype	
	Phenotype	

Make a Fairy Activity: Color & Cut Page

Directions: Color the woodland scene on page 53. Color all of your fairy's parts according to Trait 1's phenotype for coloration (either pastel or bright colors). Carefully cut out your fairy and all of your fairy's phenotypes. Glue your fairy onto the woodland scene in this order: Wings first, body second, face third, hair fourth, and shoes last.

Make a Goblin Activity

Activity Overview:

You will randomly select the traits of two goblin parents and solve Punnett squares to determine your goblin's inherited traits. Then you will color, cut, and glue your goblin together onto a cave scene you have colored.

Needed Supplies:

- Scissors
- Glue stick
- Colored pencils or crayons
- A coin or a digital coin flip
- A four-sided die or other method of randomly generating a number between one and four

Activity Directions:

1. Follow the directions on page 56 to determine the parents' genotypes and phenotypes.
2. Solve the Punnett squares on page 57 and record your goblin's genotypes and phenotypes.
3. Color the cave scene on page 61.
4. Color your goblin and all of your goblin's inherited phenotypes on page 59 according to Trait 1's phenotype for coloration (either greens/browns or reds/grays).
5. Carefully cut out the goblin and all of its phenotypes.
6. Glue your goblin's ears slightly behind its head. Glue the goblin onto the cave scene in this order: Scent clouds, goblin body (with ears attached), feet, and food (food fits into the goblin's right hand).

Make a Goblin Activity:

Parents' Genotypes & Phenotypes

Directions: Discover your goblin's parents' genotypes with coin flips. Heads is dominant and tails is recessive. To find each parent's genotype, flip a coin two times to determine what pair of alleles they have for a trait. Use the genotype to determine that parent's phenotype and record the genetic information in the table below. *Example: If your first coin flip is heads (dominant) and your second coin flip is tails (recessive), the genotype of the first parent would be Tt. For Trait 1, that parent's phenotype would be shades of green and brown.*

Trait 1: Coloration

- Shades of green and brown are dominant.
- Shades of red and gray are recessive.

Trait 2: Smell

- Smelling like flowers is dominant.
- Extra stinky and foul is recessive.

Trait 3: Ear Size

- Large ears are dominant.
- Small ears are recessive.

Trait 4: Feet Shape

- Sharp, clawed feet are dominant.
- Soft, hairy feet are recessive.

Trait 5: Diet

- Carnivores are dominant.
- Herbivores are recessive.

		Parent 1	Parent 2
Trait 1	Genotype		
	Phenotype		
Trait 2	Genotype		
	Phenotype		
Trait 3	Genotype		
	Phenotype		
Trait 4	Genotype		
	Phenotype		
Trait 5	Genotype		
	Phenotype		

Make a Goblin Activity:

Your Goblin's Genotypes & Phenotypes

Directions: Solve the Punnett squares for each of the five traits using the parents' genotypes you recorded on page 56. After you have solved the Punnett squares, double-check to make sure you solved correctly. Then roll a four-sided die to randomly select which quarter of the Punnett square contains your goblin's genotype. Make a new roll for each of the five traits. Use the trait information on page 56 to determine your goblin's phenotypes. Record your goblin's genetic information in the chart.

Trait 1

1	2
3	4

Trait 2

1	2
3	4

Trait 3

1	2
3	4

Trait 4

1	2
3	4

Trait 5

1	2
3	4

		Your Goblin
Trait 1	Genotype	
	Phenotype	
Trait 2	Genotype	
	Phenotype	
Trait 3	Genotype	
	Phenotype	
Trait 4	Genotype	
	Phenotype	
Trait 5	Genotype	
	Phenotype	

Make a Goblin Activity: Color & Cut Page

Directions: Color the cave scene on page 61. Color all of your goblin's parts according to Trait 1's phenotype for coloration (either greens/browns or reds/grays). Carefully cut out your goblin and all of your goblin's phenotypes. Glue your goblin's ears slightly behind its head. Glue your goblin onto the cave scene in this order: Scent clouds, goblin body (with ears attached), feet, and food (food fits into the goblin's right hand).

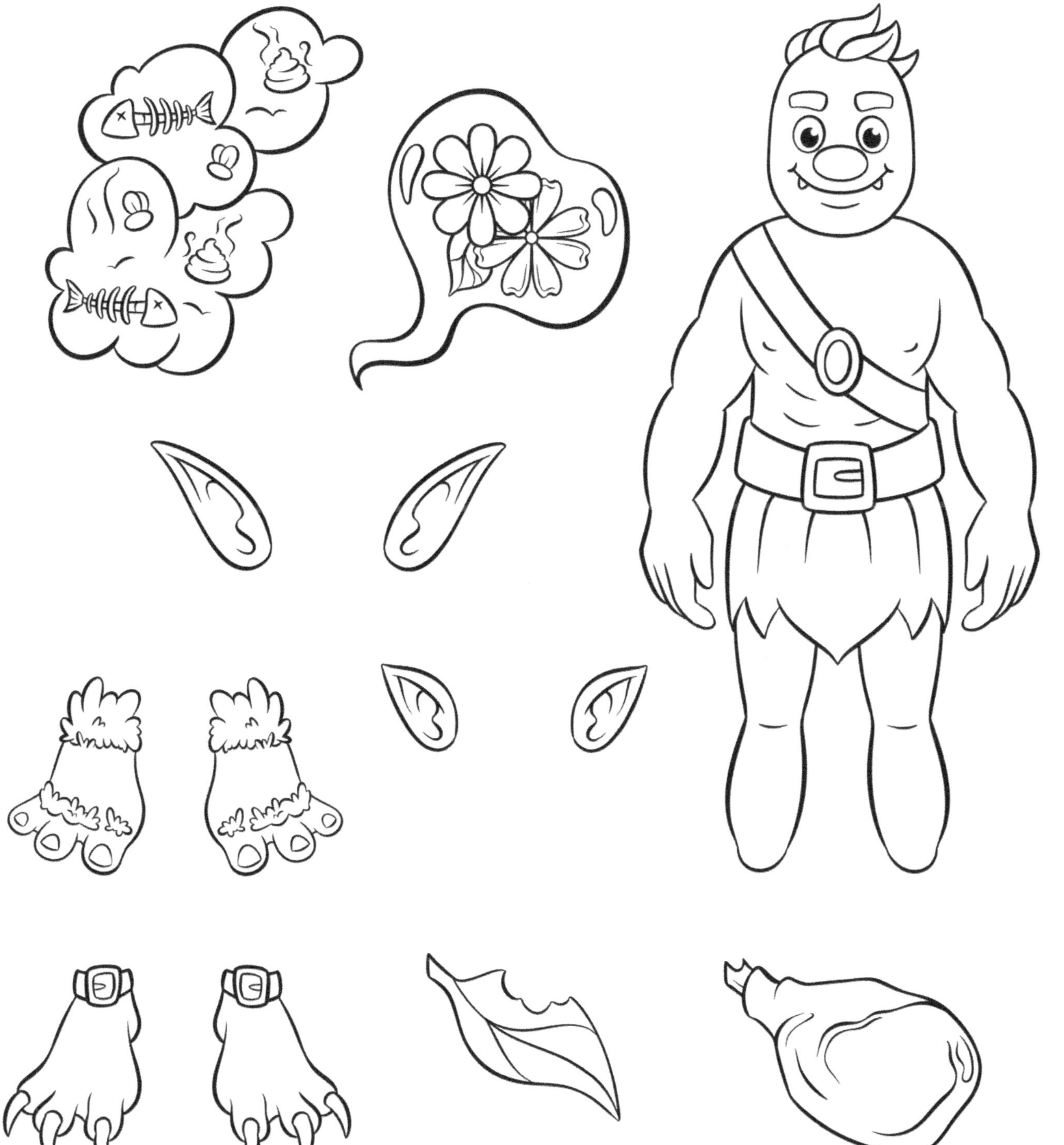

Homozygous & Heterozygous

The Same and Different

There are two words you'll see often when learning about genetics: homozygous and heterozygous. They might sound fancy and complicated, but if you break them down into their root words, they're pretty straightforward.

Homozygous can be broken down into *homo* and *zygous*. The Greek root *homo* means same, like, or equal. *Zygous* means to yoke or join together.* When you add the meaning of *homo* with the meaning of *zygous*, you get something like "same joined together," which is exactly what homozygous means. Homozygous is when two of the same alleles are joined together. A homozygous genotype could mean two dominant alleles, or it could mean two recessive alleles.

Heterozygous is just the opposite. The Greek root *hetero* means different. *Zygous* still means to join together. When you add the meaning of *hetero* with the meaning of *zygous*, you get "different joined together." Heterozygous is when two different alleles are joined together. A heterozygous genotype could mean one dominant allele and one recessive allele.

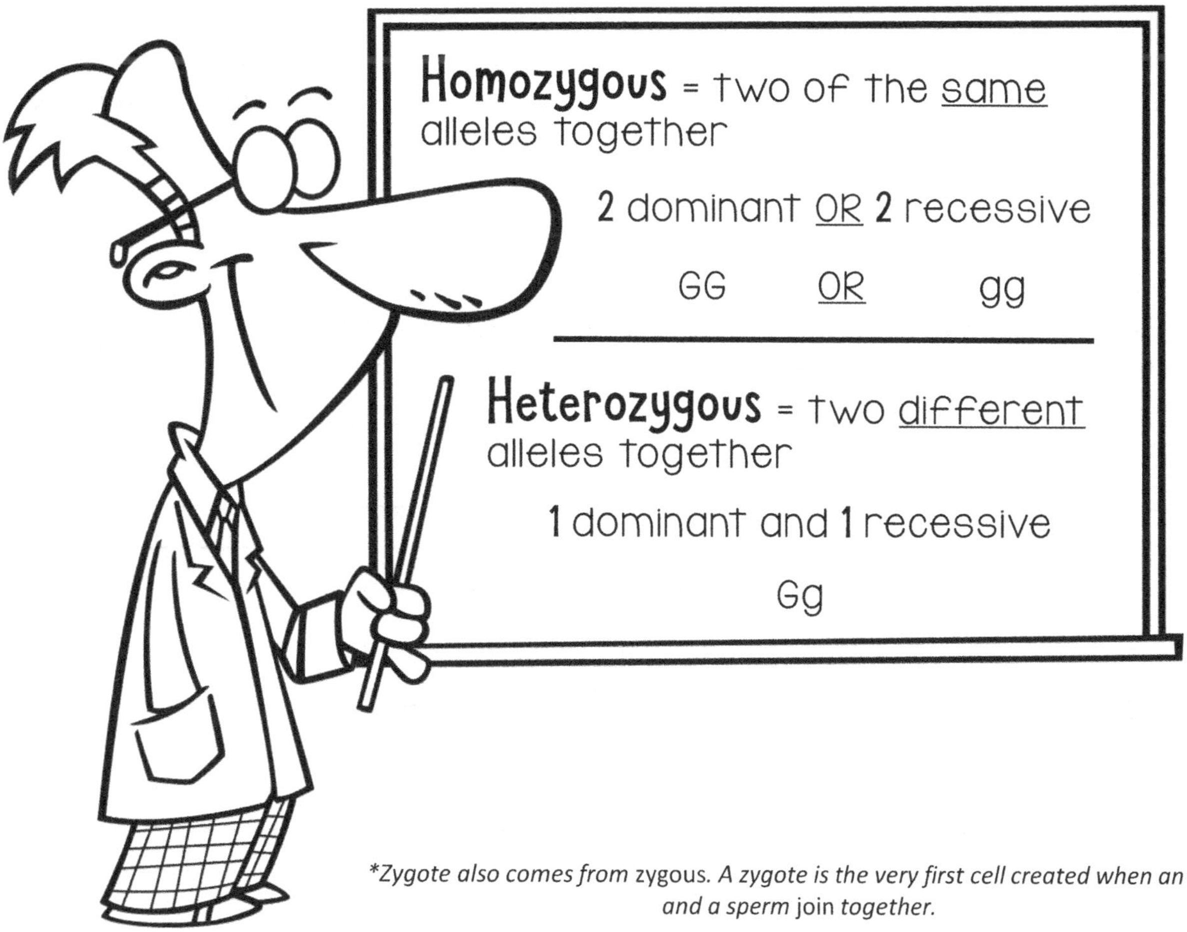

**Zygote also comes from zygous. A zygote is the very first cell created when an egg and a sperm join together.*

Homozygous & Heterozygous Practice

Directions: Answer the following questions in complete sentences.

1. In your own words, what does homozygous mean? ______________________________

2. In your own words, what does heterozygous mean? ______________________________

3. Is a pair of dominant alleles homozygous or heterozygous? How do you know? __________

4. Is a genotype with one dominant allele and one recessive allele homozygous or heterozygous? How do you know? ______________________________

5. Circle all of the examples of homozygous.

Jj	AA	uu	Two dominant alleles	A dominant paired with a recessive	Two recessive alleles

6. Circle all of the examples of heterozygous.

SS	rr	ZZ	Bb	Hybrid	Purebred

7. In order to show a recessive trait, will a genotype be homozygous, heterozygous, or either? How do you know? Explain your answer. ______________________________

8. In order to show a dominant trait, will a genotype be homozygous, heterozygous, or either? How do you know? Explain your answer. ______________________________

Homozygous & Heterozygous Corgi Coloring

Directions: Read the text beside each corgi emoji. If the text describes heterozygous, then color the corgi blue. If the text describes homozygous, then color the corgi orange. After you have colored the corgis, answer the question at the bottom of the page in complete sentences.

1

An organism has two different alleles for a trait.

2

An organism has two of the same alleles for a trait.

3

Ff

4

A purebred trait

5

Two recessive alleles are paired.

6

A hybrid trait

What is the difference between heterozygous and homozygous? ____________________

____

____

____

____

____

____

Homozygous & Heterozygous Punnett Squares

Directions: Read the questions below and circle your answers.

1. Which option below correctly represents a homozygous dominant genotype?

   HH   Hh   hh

   How can you tell? ______________________________

   ____

2. Which option below correctly represents a heterozygous genotype?

   HH   Hh   hh

   How can you tell? ______________________________

   ____

3. Which option below correctly represents a homozygous recessive genotype?

   HH   Hh   hh

   How can you tell? ______________________________

   ____

Directions: Solve each of the Punnett squares below and on the next page. Use the Punnett square results to predict the offspring. Use the capital and lowercase form of the letter E to make your Punnett squares. The first one is completed and includes detailed explanations to guide you. *Writing explanations is entirely optional.* Leave the circles empty for now.

4. **One parent is homozygous dominant. The other parent is heterozygous.**

	E	E
E	EE	EE
e	Ee	Ee

50 ___ % chance of homozygous dominant
→ Homozygous dominant is EE. The square has two homozygous dominants, which is 50%.

50 ___ % chance of heterozygous
→ Heterozygous is Ee. The square has two heterozygous pairs, which is 50%.

0 ___ % chance of homozygous recessive
→ Homozygous recessive is ee. The square has no homozygous recessive, so that is 0%.

5. **Both parents are homozygous dominant.**

______ % chance of homozygous dominant

______ % chance of heterozygous

______ % chance of homozygous recessive

Homozygous & Heterozygous Punnett Squares

6. **Both parents are homozygous recessive.**

_____ % chance of homozygous dominant

_____ % chance of heterozygous

_____ % chance of homozygous recessive

7. **Both parents are heterozygous.**

_____ % chance of homozygous dominant

_____ % chance of heterozygous

_____ % chance of homozygous recessive

8. **One parent is homozygous recessive. The other parent is heterozygous.**

_____ % chance of homozygous dominant

_____ % chance of heterozygous

_____ % chance of homozygous recessive

9. **One parent is heterozygous. The other parent is homozygous dominant.**

_____ % chance of homozygous dominant

_____ % chance of heterozygous

_____ % chance of homozygous recessive

10. **One parent is homozygous recessive. The other parent homozygous dominant.**

_____ % chance of homozygous dominant

_____ % chance of heterozygous

_____ % chance of homozygous recessive

11. **One parent is a hybrid. The other parent is a purebred for the recessive trait.**

_____ % chance of homozygous dominant

_____ % chance of heterozygous

_____ % chance of homozygous recessive

Homozygous & Heterozygous Punnett Squares

Now it's time to fill in those empty circles on pages 66-67.

Directions: In each of the empty circles, you will answer the following question:

What percent of the offspring are expected to show the dominant trait?

Use these hints to help you answer the question above:

- The dominant trait will be seen if a dominant allele is present. There are two ways to see a dominant trait.
    - A homozygous dominant genotype (example TT) will show a dominant trait.
    - A heterozygous genotype (example Tt) will show a dominant trait.
- A recessive trait will only be seen if both alleles are recessive.
    - A homozygous recessive genotype (example tt) will show a recessive trait.
- To determine the percent of offspring expected to show the dominant trait, you need to look at the Punnett square results. How many of the genotypes inside of the Punnett square have one or more dominant alleles?
    - If all of the genotypes have at least one dominant allele, that is four out of four or 100%.
    - If three of the genotypes have at least one dominant allele, that is three out of four or 75%.
    - If two of the genotypes have at least one dominant allele, that is two out of four or 50%.
    - If one of the genotypes has at least one dominant allele, that is one out of four or 25%.
    - If none of the genotypes has at least one dominant allele, that is zero out of four or 0%.

An Example:

To the right is the completed example from page 66. Examine the inside of the Punnett square, and notice all four quarters have either two dominant alleles (EE) or one dominant allele (Ee). Both EE and Ee will show a dominant trait. Therefore, four out of four or 100% of the offspring are expected to show the dominant trait.

Directions: Go back to pages 66-67 and correctly fill in the circles to show the percent of offspring expected to show the dominant trait.

4. **One parent is homozygous dominant. The other parent is heterozygous.**

	E	E
E	EE	EE
e	Ee	Ee

50___ % chance of homozygous dominant
→ *Homozygous dominant is EE. The square has two homozygous dominants, which is 50%.*

50___ % chance of heterozygous
→ *Heterozygous is Ee. The square has two heterozygous pairs, which is 50%.*

0___ % chance of homozygous recessive
→ *Homozygous recessive is ee. The square has no homozygous recessive, so that is 0%.*

100%

This is the completed example from page 66 showing the percent of offspring expected to show the dominant trait.

Colorful Monster Phenotypes with Provided Genotypes

Directions: Read each question and solve the Punnett square to find each monster's genotype. Then color the monster's phenotype accordingly. Color the rest of the monster however you want.

1. Heela's mother has a lovely, yellow star. Her father is heterozygous for his orange star. Orange stars are dominant and yellow stars are recessive. What is Heela's genotype and phenotype?

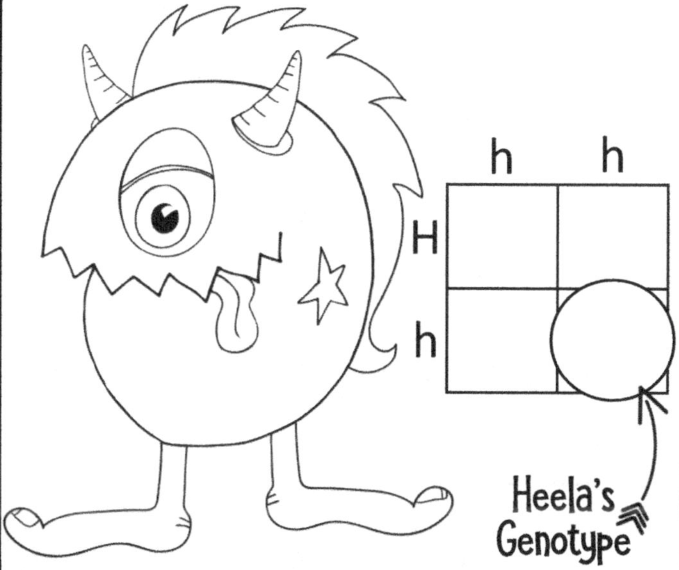

2. Roary's father comes from a long, purebred line of red tails. His mother also has a red tail, but she is heterozygous for her coloration. Red tails are dominant over brown tails. What is Roary's genotype and phenotype?

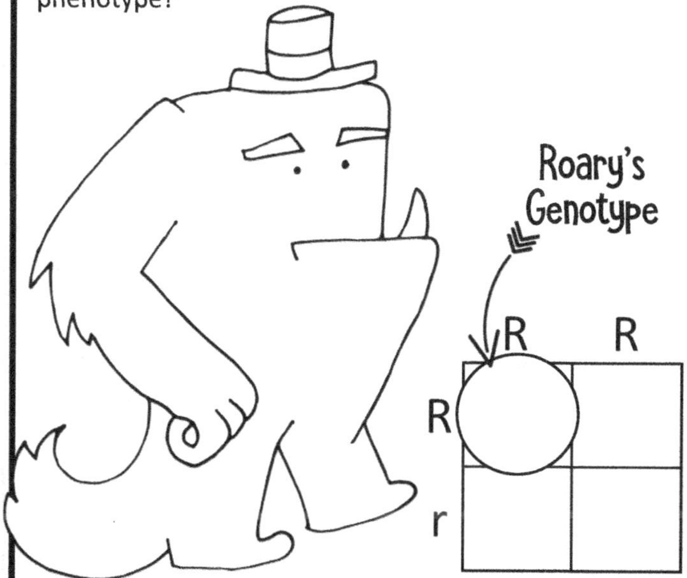

3. In Olly's family, orange and green spots are dominant over purple and blue spots. Olly's mom is homozygous for her orange and green spots, and his father is homozygous for his purple and blue spots. What is Olly's genotype and phenotype?

4. Jeepers' mom has fuzzy green hair and is heterozygous for this trait. His dad has shaggy blue hair. Green hair is dominant over blue hair. What is Jeepers' genotype and phenotype?

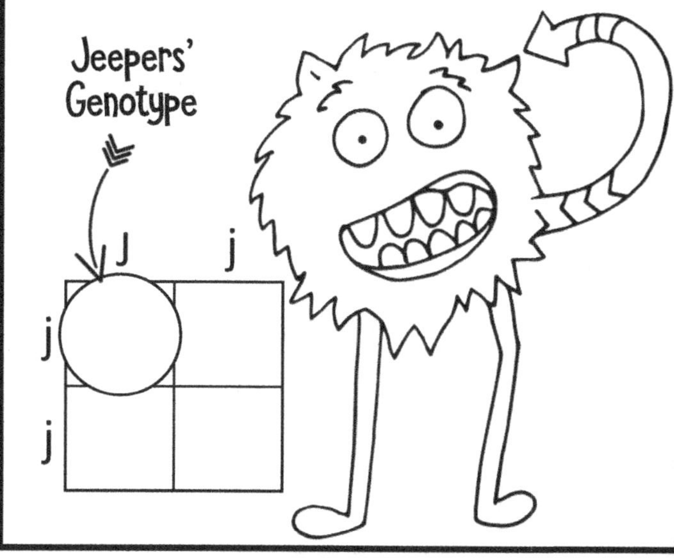

Colorful Monster Phenotypes with Genotype Clues

Directions: Read each question and solve the Punnett square to find each monster's genotype. Then color the monster's phenotype accordingly. Color the rest of the monster however you want.

1. UggUgg's parents both have pink tusks, which are dominant over red tusks. However, one of her parents is homozygous and the other is heterozygous. What is UggUgg's genotype and phenotype?

2. In Gargh's family, orange snail eyes are recessive and considered gorgeous. Red snail eyes are dominant but hideous. His mother has beautiful orange eyes. His father has heterozygous red eyes. What is Gargh's genotype and phenotype?

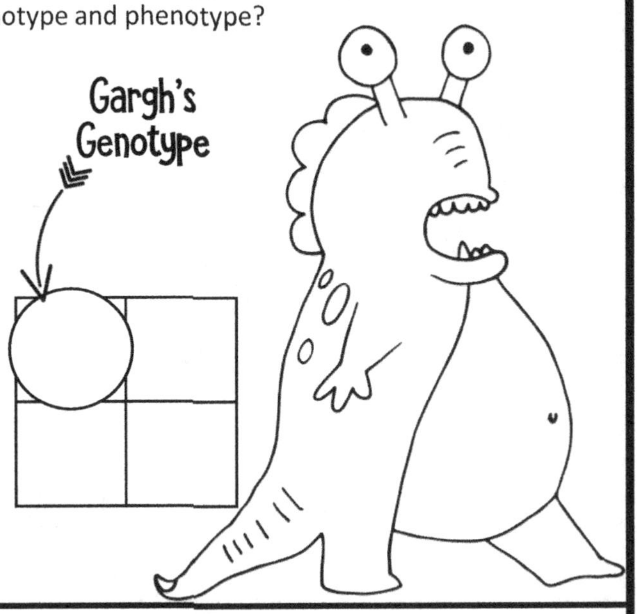

3. In Steamer's family, brown eyebrows are dominant over yellow eyebrows. Steamer's dad is a purebred for his brown eyebrows. His mom also has brown eyebrows but has a different genotype. What is Steamer's genotype and phenotype?

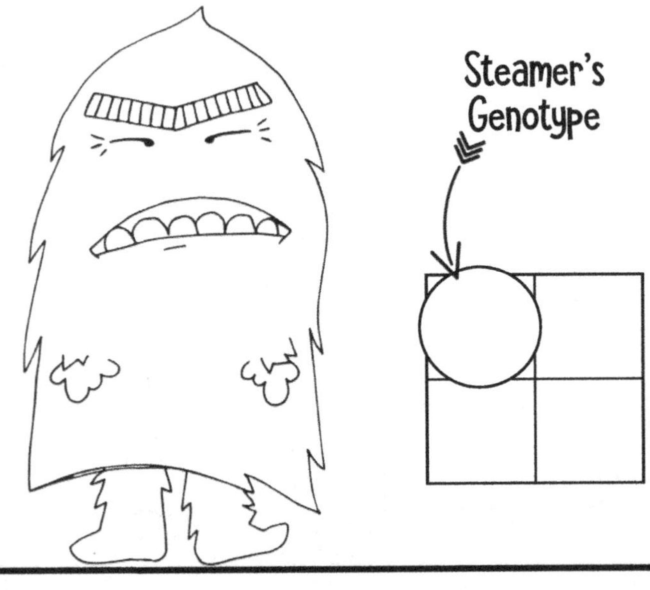

4. Some of Eyela's older brothers and sisters have dominant blue tongues while others have recessive green tongues. Eyela's parents both have blue tongues and share the same genotype. What is Eyela's genotype and phenotype?

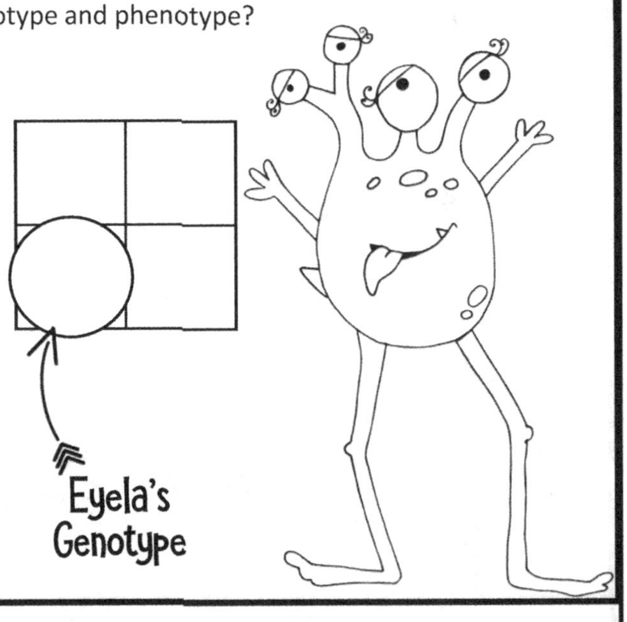

Make Your Own Colorful Monsters

Directions: Find each monster's genotype and phenotype by choosing its parents' genetic makeup. Solve the Punnett square, and color the monster accordingly. Color the rest of the monster however you want.

1. Dominant Phenotype (circle one):

red feet   green feet   brown feet   yellow feet

Recessive Phenotype (circle one):

blue feet   purple feet   orange feet   pink feet

Mother's Genotype (circle one):

homozygous dominant   homozygous recessive   heterozygous

Father's Genotype (circle one):

homozygous dominant   homozygous recessive   heterozygous

Monster's Name: ______________________________

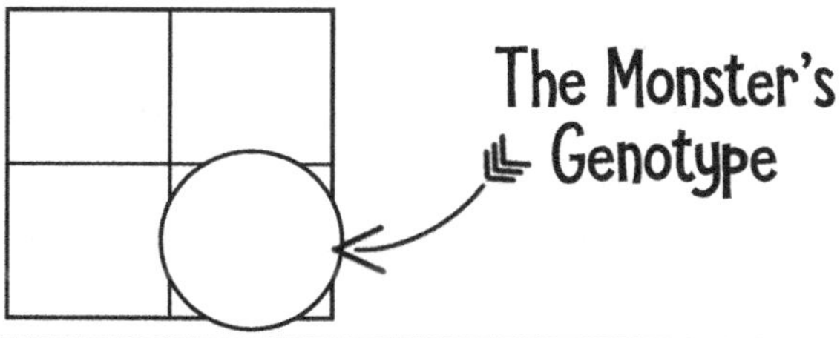

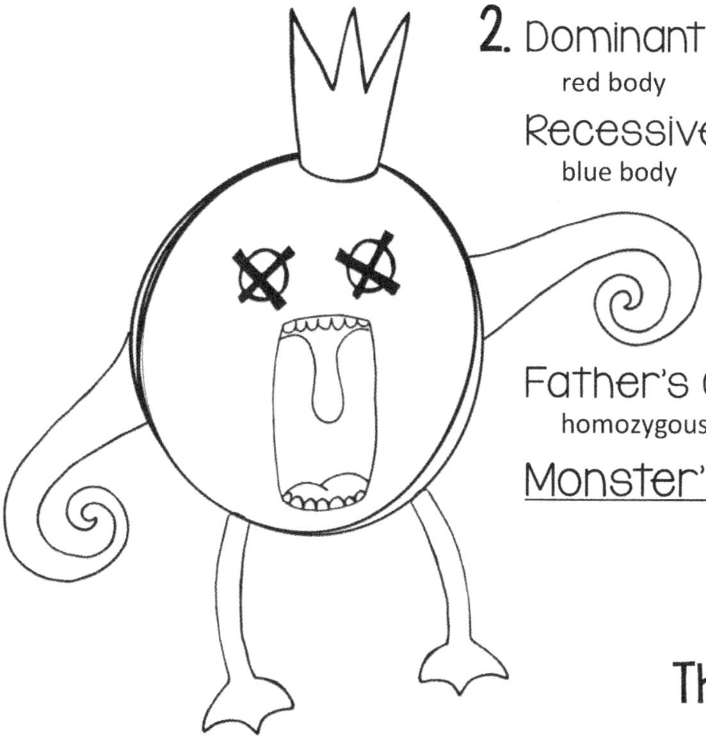

2. Dominant Phenotype (circle one):

red body   green body   brown body   yellow body

Recessive Phenotype (circle one):

blue body   purple body   orange body   pink body

Mother's Genotype (circle one):

homozygous dominant
homozygous recessive
heterozygous

Father's Genotype (circle one):

homozygous dominant   homozygous recessive   heterozygous

Monster's Name: ______________________________

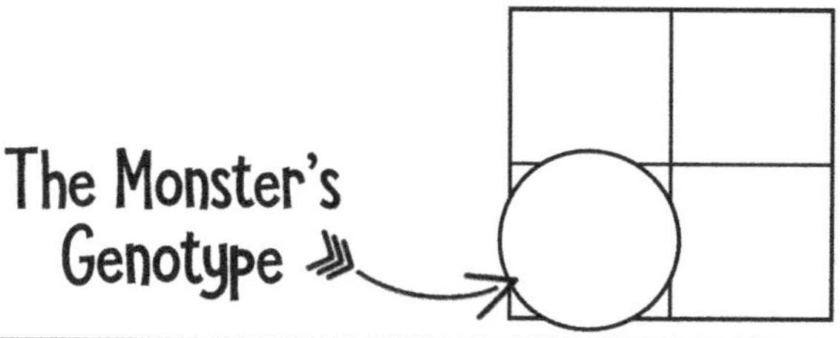

Homozygous & Heterozygous Experiment

Experiment Overview:

In this experiment you will randomly generate forty genotypes (pairs of alleles). Then you will analyze the percents of homozygous and heterozygous genotypes. You will also determine how many of the genotypes will show the dominant trait and how many will show the recessive trait.

How to Make Allele Pairs:

In order to generate your genotypes, you need a coin to toss.* Heads represents a dominant allele (T). Tails represents a recessive allele (t). To make each genotype, you will toss the coin twice. For example, if your first toss is heads and your second toss is tails then your genotype is Tt. Alternatively, if both tosses are heads, then your genotype is TT.

Experiment Directions:

- Read this entire set of directions before beginning.
- Complete the "Your Predictions" section below.
- Make coin tosses to create your genotypes. You need forty genotypes. That means you will toss your coin a total of eighty times. You must record all of your genotypes on the next page in the "Your Genotypes" chart. Remember to use T for dominant alleles and t for recessive alleles.
- Analyze your results and reflect on the accuracy of your predictions.

Your Predictions:

Record and explain your predictions in the chart below.

What percent of your genotypes will be <u>homozygous</u>?	**What percent of your genotypes will be <u>heterozygous</u>?**	**What percent of your genotypes will show the <u>dominant phenotype</u>?**	**What percent of your genotypes will show the <u>recessive phenotype</u>?**
%	%	%	%
Explain the reasoning behind your homozygous and heterozygous predictions.		Explain the reasoning behind your dominant and recessive predictions.	

**You can use something other than a coin if you prefer. You can roll a six-sided die and use odd numbers to represent dominant alleles and even numbers to represent recessive alleles. You can place two differently colored pompoms in a paper bag and draw them at random, with one color representing dominant and the other recessive. You can find a digital coin toss online.*

Homozygous & Heterozygous Experiment

Your Genotypes	
1.	21.
2.	22.
3.	23.
4.	24.
5.	25.
6.	26.
7.	27.
8.	28.
9.	29.
10.	30.
11.	31.
12.	32.
13.	33.
14.	34.
15.	35.
16.	36.
17.	37.
18.	38.
19.	39.
20.	40.

After You've Recorded Your Genotypes:

Once you have recorded all forty of your genotypes in the chart on the left, complete each of the sections below.

Homozygous

Your homozygous genotypes are either TT or tt. How many of your genotypes are homozygous?

Divide that number by 40 and multiply the resulting quotient by 100 to determine the percent of homozygous genotypes.* Record that percent.

How close was your prediction?

Heterozygous

Your heterozygous genotypes are Tt. How many of your genotypes are heterozygous?

Divide that number by 40 and multiply the resulting quotient by 100 to determine the percent of heterozygous genotypes. Record that percent.

How close was your prediction?

Dominant & Recessive Traits

A dominant trait is seen with TT and Tt. How many of your genotypes will show the dominant trait? What percent is this?

A recessive trait is seen with tt. How many of your genotypes will show the recessive trait? What percent is this?

How close were your predictions?

**Example: Pretend you have 36 homozygous genotypes.*
36 ÷ 40 = 0.9 → 0.9 x 100 = 90   That means 90% are homozygous.

Homozygous & Heterozygous Experiment

Summary of the Results:

Write a paragraph of five or more sentences to summarize the results of the experiment. Use the guiding thoughts and questions below to help you.

- Compare the number of homozygous genotypes to the number of heterozygous genotypes.
- What percent of the genotypes will show the dominant trait?
- Compare the numbers of homozygous dominant to homozygous recessive.
- What trends or patterns do you see in the numbers?

Homozygous & Heterozygous Experiment

Repeating the Results...Or Not?

Think about the results *you* got from this experiment. If a friend did this same experiment, would their results be the same as yours? Why or why not? Would a larger sample size of 1,000 genotypes have results similar to yours? What about a smaller sample size of only 10? Explain your answers in sentences.

Bringing Back the Pea Dude:

Think about the results you got from this experiment. How are your results similar to Gregor Mendel's pea plant studies? Draw comparisons from your results to Mendel's experiments. How are they similar? How are they different? Do you think this experiment accurately represents Mendel's studies? Why or why not? Explain your answers in sentences.

Famous Monster Genetics

Directions: Solve the genetic questions using a Punnett square. Use capital and lowercase letter B for your Punnett squares. Then complete each table with the possible genotypes, phenotypes, and the percent chance of each offspring type. The first one is completed for you.

1. A certain giant lizard monster and his partner are thinking about starting a family. They both have the same heterozygous genotype for their green scales. They want to know if they can have baby lizards with rare recessive black scales.

	B	b
B	BB	Bb
b	Bb	bb

Genotype	Phenotype	Percent Chance
BB	Green scales	1 out of 4 = 25% chance
Bb	Green scales	2 out of 4 = 50% chance
bb	Black scales	1 out of 4 = 25% chance

2. The Abominable Snowman has beautiful white fur that makes him difficult to see in the snowy mountains. White fur is dominant in yetis, and the Abominable Snowman is heterozygous for this trait. His partner, the Abhorrent Slushlady, is a glimmering gray color because of her recessive homozygous genotype. What kind of fur are their offspring likely to inherit?

Genotype	Phenotype	Percent Chance
BB		
Bb		
bb		

3. Frankenstein and his Bride are expecting a little one. Frankenstein is a purebred for his recessive smooth hair. Frankenstein's Bride is a hybrid for her dominant wiry hair. They want to know what kind of hair their baby is most likely to inherit.

Genotype	Phenotype	Percent Chance
BB		
Bb		
bb		

Famous Monster Genetics

4. Nessie the Loch Ness Monster was born very shy, which is a dominant trait in lake monsters. Her fiancé, Nester, is also shy. Nessie and Nester are both heterozygous for their shyness. They want to know if their future offspring have any chance of being outgoing.

Genotype	Phenotype	Percent Chance
BB		
Bb		
bb		

5. When King Kong was taken away from his home on Skull Island, his mate was expecting a baby. King Kong always wondered if his baby inherited the same recessive homozygous genotype that caused him to grow into such a large adult. His mate was homozygous for her average size. Help King Kong learn the possible sizes of his baby.

Genotype	Phenotype	Percent Chance
BB		
Bb		
bb		

6. Bigfoot recently met the love of his life, Microfoot. The only problem is that she has tiny feet, which is an undesirable recessive trait among their kind. Bigfoot is a purebred for his large feet. Help them find out if their future babies will have feet like their father or their mother.

Genotype	Phenotype	Percent Chance
BB		
Bb		
bb		

Famous Monster Genetics

Directions: Choose one of the famous monster families from questions 1-6 on pages 76-77. In the space below, draw the parents with an offspring who shows the trait most likely to be inherited.

The Monster Family

Homozygous & Heterozygous Assessment

1. ______________________________ refers to having two different alleles for a trait.

2. ______________________________ refers to having two of the same alleles for a trait.

3. True or False: GG and Gg are both examples of a heterozygous genotype.

4. True or False: Homozygous genotypes can only have dominant alleles.

For questions 5-8, underline the heterozygous genotypes and circle the homozygous genotypes.

5. TT

6. hh

7. Ww

8. Yy

Answer question 9 using complete sentences.

9. A purebred comes from a family with many generations of the same trait. Are purebreds heterozygous or homozygous? Explain your answer.

____

____

____

____

____

____

____

____

____

____

____

Review of Genetics Concepts

Directions: Draw a line from the concept on the left to its description on the right.

Alleles	It is the scientific study of heredity.
Dominant	It is when two different alleles are paired together.
Genes	It is the allele combination or genetic makeup of an organism.
Genetics	It is when two of the same alleles are paired together.
Heredity	It is the one whose trait always shows up if the allele is present.
Heterozygous	They are the characteristics or features of an organism.
Homozygous	It is the passing of traits from parents to offspring.
Hybrid	They are the different forms of the genes. They might be dominant or recessive.
Gregor Mendel	It's the dominant phenotype!
Genotype	It is an offspring with two distinctly different parental types.
The Principle of Dominance	It's the recessive phenotype!
Purebred	Its trait only shows up when no dominant alleles are present.
Recessive	They are the basic units of heredity that contain information about the traits passed from parents to offspring.
Traits	He is known as the father of modern genetics. No, you may not call him the Mendelorian.
It's seen when two recessive alleles pair.	It is an organism with many generations of the same trait.
It's seen when one dominant and recessive allele pair.	When an organism has two different alleles, the dominant trait will be expressed and the recessive trait will be masked.

Dominant & Recessive Dog Traits

Directions: Use the data table to answer the questions that follow.

Traits in Dogs	
Dominant Traits	**Recessive Traits**
Brown eyes (B)	Blue eyes (b)
Short coat (S)	Long coat (s)
Wiry fur (W)	Smooth fur (w)
Curly fur (C)	Straight fur (c)
Large ears (L)	Small ears (l)
Stickup ears (E)	Dropped ears (e)

1. A dog has a genotype of ee. What is his phenotype? ______________________________

2. A dog has a genotype of WW. What is her phenotype? ______________________________

3. A dog has a genotype of Bb. What is her phenotype? ______________________________

4. Can a dog with a long coat be heterozygous, homozygous, or both? ____________________

5. A dog has purebred dominant stickup ears. What is his written genotype? ____________

6. What is the genotype of a dog heterozygous for brown eyes? _________________________

7. What genotype(s) is possible for a dog with large ears? ______________________________

8. What genotype(s) is possible for a dog with straight fur? ____________________________

9. Fluffy and Fido are having a litter of puppies soon! Fluffy is purebred dominant for curly fur. Fido has straight fur. Complete the Punnett square. What are the chances of their puppies having straight fur?

____

____

10. Buddy and Babe are both heterozygous for large ears. What kind of ears are their puppies most likely to have? Complete the Punnett square and then answer the two questions below.

a. How likely are their puppies to have the same genotype as their parents? __________

b. How likely are their puppies to have the same phenotype as their parents? __________

Genetics Vocabulary Review Volume 1

Directions: For each genetics vocabulary word, use the word correctly in a sentence and draw a picture to help you remember the definition. The definitions are provided for you.

Heredity: The passing of characteristics from parents to offspring Sentence: Picture:	Genetics: The scientific study of heredity Sentence: Picture:
Traits: The characteristics that are passed from parents to offspring Sentence: Picture:	Genes: The basic units of heredity that contain instructions about which traits an organism will develop Sentence: Picture:
Alleles: The different forms of a gene Sentence: Picture:	Punnett Square: A square used to calculate the frequences of genotypes and phenotypes among offspring Sentence: Picture:

Genetics Vocabulary Review Volume 2

Directions: For each genetics vocabulary word, provide an example and draw a picture to help you remember the definition. The definitions are provided for you.

Genotype: The genetic makeup or allele combination of an organism Example: Picture:	Phenotype: The observable traits seen in an organism Example: Picture:
Homozygous: An organism with two of the same alleles for a trait Example: Picture:	Heterozygous: An organism with two different alleles for a trait Example: Picture:
Purebred: An organism with many generations of the same trait Example: Picture:	Hybrid: The offspring of two distinctly different parental types Example: Picture:

Farm Animal Genetics

Directions: Use the Dominant and Recessive Traits in Farm Animals table to answer questions 1-10.

Dominant & Recessive Traits in Farm Animals		
Animal	Dominant Trait	Recessive Trait
Cats	Short hair (M)	Long hair (m)
Cattle	Black hair (B)	Red hair (b)
Chickens	Feathered shanks (F)	Clean shanks (f)
Dogs	Solid color (D)	White spotting (d)
Ducks	Colorful feathers (A)	White feathers (a)
Horses	Smooth hair (H)	Curly hair (h)
Pigs	Upright ears (E)	Drooping ears (e)
Sheep	Brown eyes (L)	Light eyes (l)

1. A dog is purebred for his white spotting. What is the dog's genotype?

 ____

2. A sheep's genotype is Ll. What is its phenotype?

 ____

3. A cow has red hair. What is its genotype? What is its phenotype?
 Genotype: ______________________________
 Phenotype: ______________________________

4. One chicken has an FF genotype. Another chicken has an Ff genotype. Do they share the same phenotype? ______________

5. In horses, what genotype(s) will display smooth hair?

 ____

6. What is the phenotype of a duck with an Aa genotype?

 ____

7. A sheep has light eyes. What is its genotype?

 ____

8. A dog is homozygous for its solid color. What is the dog's genotype?

 ____

9. What is the genotype and phenotype of a heterozygous horse?
 Genotype: ______________________________
 Phenotype: ______________________________

10. In cats, what genotype(s) will display the dominant phenotype?

 ____

Farm Animal Genetics

Directions: Use the Dominant and Recessive Traits in Farm Animals table on the previous page to complete the Punnett squares. Then answer questions 11-17.

11. Two white ducks are crossed. What percent of their ducklings are predicted to have white feathers? Write your answer in the box.

12. A pair of heterozygous cats is expecting kittens. Complete a pie chart to show the percent chance of each possible genotype of their kittens.

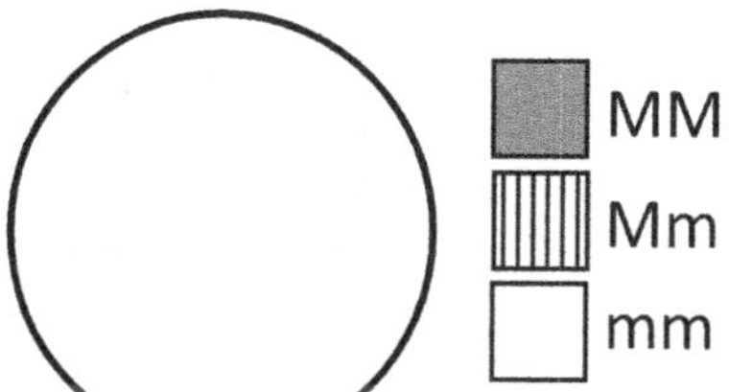

13. A stallion with a pair of dominant alleles is crossed with a mare with one dominant and one recessive allele. Their filly is born with smooth hair. What is the likelihood her genotype is homozygous? Write your answer in the box.

14. A cow with black hair and a heterozygous genotype is crossed with a bull with red hair. What are the chances their calf will be born displaying the recessive trait? Write your answer in the box.

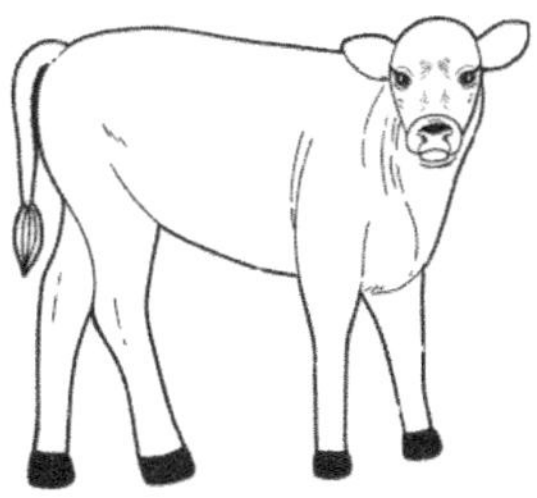

Farm Animal Genetics

15. A sow and a boar both have upright ears, but they have different genotypes. Complete the pie chart to show the percent chance of each possible genotype of their piglets. What percent of their piglets are expected to have a heterozygous genotype? Write your answer in the box.

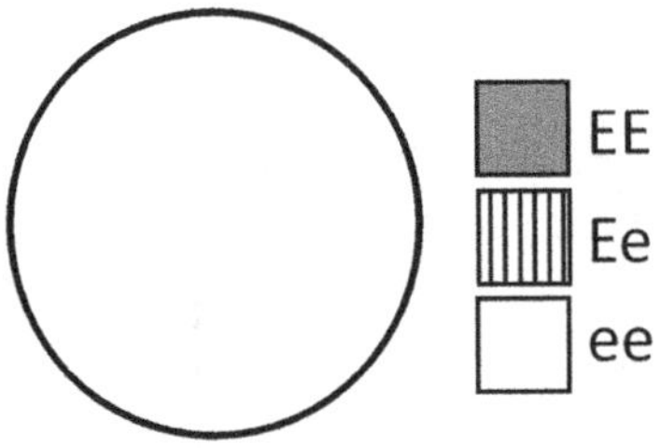

16. A purebred hen with feathered shanks is crossed with a purebred rooster with clean shanks. What percent of their chicks are expected to have a homozygous genotype?

17. Create and solve your own genetics question using the Dominant and Recessive Traits in Farm Animals table on page 84.

Make an Organism Project

Project Overview & Options:

For this project you will select an organism, choose multiple dominant and recessive traits, determine the parents' genotypes and phenotypes, and create their offspring. Your organism can be a real-life living thing, a mythical creature, or a living thing entirely from your imagination.

- Option 1: Make a physical model of your offspring to show its inherited phenotypes. You can create your model using clay, popsicle sticks, beads, construction paper, pipe cleaners, leaves, and/or other materials you have on hand.
- Option 2: Make a digital model of your offspring to show its inherited phenotypes using AI to generate an image.
- Option 3: Draw and color your offspring on paper to show its inherited phenotypes.

Needed Supplies:

- A coin or digital coin flip
- A four-sided die or other method of randomly generating a number between one and four
- Any materials needed to create your offspring model

Project Directions:

An overview of the steps is described below. The detailed example on page 88 can be used to guide you. The student project pages on 89-90 also list the steps.

1. Decide what kind of organism you will make on page 89.
2. Determine six different traits and their dominant and recessive versions. Record that information in the chart on page 89.
3. Using coin tosses, determine both parents' genotypes and phenotypes. A parent's genotype is determined with two coin tosses. Heads represents a dominant allele (T). Tails represents a recessive allele (t). *(Example: Both coin tosses are tails, so a parent's genotype for a trait is tt.)* Record the parents' genotypes and phenotypes in the parent chart on page 89.
4. Add the parents' genotypes to the Punnett squares and solve to determine the possible traits of their offspring on page 90.
5. For each of the six traits, roll a four-sided die to discover which quarter of the Punnett square has your organism's genetic information. A four-sided die will be rolled a total of six times, once for each trait. *(Example: A two was rolled for Trait 1, so quarter two of Trait 1's Punnett square gets circled. The genotype in that quarter is used to determine the organism's phenotype.)* Record your organism's inherited genotypes and phenotypes in the chart on page 90.
6. Create a model of your organism using Option 1, 2, or 3 above to show all of its inherited phenotypes.
7. Answer the questions about your organism on page 91.

Make an Organism Project Example

My Organism:
Monster

Decide what your organism will be and write it in the corresponding box on page 89.

This page shows a completed organism example to demonstrate expectations and to help guide you through creating your own.

Choose six different traits and write them in the first column. Choose the dominant and recessive versions of each trait and record them in the appropriate columns.

Trait	Dominant Trait (T)	Recessive Trait (t)
1. Tail	Spikes	No spikes
2. Toes	3 toes	2 toes
3. Teeth	Pointy	Rounded
4. Skin color	Gray	Red
5. Personality	Mean	Friendly
6. Height	Short	Tall

Grab a coin and do two coin flips to determine Parent 1's Trait 1 genotype. Do two more coin flips to determine Parent 2's Trait 1 genotype. Use their genotypes to determine their phenotypes. Do more pairs of coin flips to determine all six of each parents' traits.*

Example: Trait 1 is about tail type. The two coin flips for Parent 1 were one heads and one tails. That means Parent 1 got one dominant (T) and one recessive (t) allele to make a genotype of Tt. Since the dominant trait is spikes and a dominant allele is present in the genotype, Parent 1's phenotype is spikes. The two coin flips for Parent 2 were both tails. That means Parent 2 got two recessive alleles for a genotype of tt. Since the recessive trait is no spikes and there are no dominant alleles, Parent 2's phenotype is no spikes.

**Remember, heads is dominant and tails is recessive.*

Trait	Parent 1	Parent 2
1. Tail	Tt: Spikes	tt: No spikes
2. Toes	tt: 2 toes	Tt: 3 toes
3. Teeth	Tt: Pointy	tt: Rounded
4. Skin color	TT: Gray	TT: Gray
5. Personality	Tt: Mean	Tt: Mean
6. Height	TT: Short	Tt: Short

Add the genotypes of Parent 1 and 2 to each of the appropriate Punnett squares and solve. Trait 1's Punnett square is shown below. Parent 1's genotype of Tt is on top and Parent 2's genotype of tt is on the left side of the Punnett square.

Trait 1:

	T	t
t	Tt (1)	tt (2)
t	Tt (3) (circled)	tt (4)

After solving the Punnett squares, roll a four-sided die to determine which quarter of the Punnett square contains the organism's genotype. In the example on this page, a 3 was rolled so quarter 3 was circled. The monster's genotype is Tt for this trait, so its phenotype is a spiked tail. That information was recorded in the My Organism's Traits table. A four-sided die will be rolled a total of six times, once for each trait. Make your model once all genotypes and phenotypes are identified.

My Organism's Traits

Trait	Genotype	Phenotype
1.	Tt	Spiked tail

Make an Organism Project

My Organism:

When doing your coin tosses, heads is dominant (T) and tails is recessive (t).

Trait	Dominant Trait (T)	Recessive Trait (t)
1.		
2.		
3.		
4.		
5.		
6.		

Trait	Parent 1	Parent 2
1.		
2.		
3.		
4.		
5.		
6.		

Project Directions:

1. Decide what organism you will make and write it in the *My Organism* box on this page.
2. Choose six traits and their dominant and recessive versions. Record that information in the chart above.
3. Using coin tosses, determine both parents' genotypes and phenotypes. Record that information in the parent chart to the left.
4. Add the parents' genotypes to the Punnett squares on page 90 and solve to determine the possible traits of their offspring.
5. Roll a four-sided die to discover which quarter of the Punnett square contains your organism's genetic information. Record your organism's inherited genotypes and phenotypes in the chart on page 90.
6. Create your offspring model.
7. Answer the questions on page 91.

Make an Organism Project

Trait 1:

1	2
3	4

Trait 2:

1	2
3	4

Trait 3:

1	2
3	4

Trait 4:

1	2
3	4

Trait 5:

1	2
3	4

Trait 6:

1	2
3	4

Directions:

Once you have completed all of the above Punnett squares, you need to determine what genetic information will be inherited by your organism.

1. Start with the first trait. Roll your four-sided die one time.
2. The number you rolled corresponds to the quarter of the Punnett square containing your organism's genotype. Circle that quarter.
3. In the chart on the right next to the appropriate trait, record the circled genotype and write in the associated phenotype.
4. Repeat steps 1-3 for the remaining traits.

My Organism's Traits

Trait	Genotype	Phenotype
1.		
2.		
3.		
4.		
5.		
6.		

Make an Organism Project

Project Reflection Questions:

1. How many recessive phenotypes did your organism inherit? ______________________

2. How many of your organism's traits were heterozygous? ______________________

3. Was your organism a purebred for any trait(s), with both parents sharing the same homozygous genotype? ______________________

4. If your organism's parents had more offspring, would they all look like your organism? Why or why not? Explain your answer in sentences. ______________________

Beyond Just Dominant & Recessive

Humans Are Not Peas

In pea plants, a tall parent crossed with a short parent never results in a medium-sized offspring. Obviously, that's not the case with people because your freakishly tall Uncle Archibald and your petite Aunt Jamilah gave you ten cousins of varying heights. You know about dominant and recessive alleles, but they alone cannot explain the variation seen among people.

Some organisms, like pea plants, are simple. Many of a pea plants' traits are monogenic, meaning a trait (such as height) is controlled by *one* single gene that might come in dominant (tall) and recessive (short). Other organisms, like people, are incredibly complex. Many of our traits are polygenic, meaning a trait like height is controlled by *multiple* genes. Each of those genes makes an impact on a person's overall height. Even a trait that seems simple, like human eye color, is not straightforward. Our exact eye color is believed to involve as many as *sixteen* different genes working together.

Monogenic: Monogenic means a trait is controlled by just <u>one</u> gene.

Pea plant flower color is monogenic with dominant being purple and recessive being white.

Polygenic: Polygenic means a trait is controlled by <u>multiple</u> genes working together.

A person's skin color is polygenic, which is why we can be very dark, very pale, and all sorts of shades in between.

Having many polygenic traits is not the only reason for people being more complex than pea plants. Sometimes people have the right genes for a trait, but that trait won't show up unless certain environmental factors are present. Allergies are a great example of this. A person might be genetically predisposed to developing allergies, but those allergies will only develop after exposure to certain allergens. Along those same lines, much of a person's development occurs as a fetus during pregnancy. What is eaten during that pregnancy and both parents' exposures to certain environmental toxins can impact which traits are seen after birth. The genes may be there—but they might be expressed differently depending on what is going on around the person during and after development.

But wait! There are even *more* reasons why people are so varied! Mutations, codominance, incomplete dominance, and acquired traits add interesting diversity. Heredity is complicated.

Beyond Just Dominant & Recessive Practice

Directions: Answer the following questions in complete sentences.

1. In your own words, what is monogenic? ______________________________

2. In your own words, what is polygenic? ______________________________

3. Why do people come in different heights but pea plants only come in tall or short? ______

4. Using what you learned about polygenic traits, list five human traits you think are probably polygenic. Why do you think these traits are most likely polygenic and not monogenic? ______________________________

5. Why might a person who is genetically predisposed to allergies never develop an allergy?

Getting Your Traits

Why are you the way you are? Is it because of your genetic makeup or your life experiences?

- In the space below, use sentences and/or pictures to answer the questions above. Be detailed and use specific examples about yourself.

Inherited & Acquired Traits

How You Got Your Traits

As you know, traits are the characteristics passed down from parents to offspring. Everyone has a different combination of traits that makes them different from all the other people in the world. There are two ways to get these traits.

The first way to get traits is genetically from parents; this is called an inherited trait. Gregor Mendel focused on inherited traits like height and flower color in his pea plants. Examples of inherited traits in people are having bushy eyebrows, long eye lashes, a wide nose, and stubby toes.

The second way to get traits is through experiences. These traits are called acquired or learned traits. Your family and your culture, along with where and how you live can influence acquired traits. Examples of acquired traits are loving salty popcorn, being able to eat with chopsticks, being terrified of clowns, and having a scar.

Inherited Trait: An inherited trait is a trait passed down genetically from parents to offspring.

Acquired Trait: An acquired trait, also called a learned trait, is one that develops through experiences or environmental factors.

An Incredible Athlete's Traits

Picture the fastest swimmer in the world. This swimmer's combination of inherited and acquired traits aids him in setting world records. His overall height and the width of his shoulders certainly makes an impact on his speed, but so does his strong sense of competition and drive to win. The access he has to state-of-the-art swimming facilities and the best coaches assist him in developing the perfect swimming technique. Spending time practicing and lifting weights impacts the size and strength of his muscles. There are countless other traits playing small and large parts in creating this record-breaking athlete.

Directions: Choose three colors of pencils or highlighters. Underline or highlight the swimmer's inherited traits in one color, the acquired traits in a second color, and traits that are likely a mix of inherited and acquired in a third color.

Inherited & Acquired Traits Practice

Directions: Answer the following questions in complete sentences.

1. What is an inherited trait? Put the definition in your own words. ______________________
____
____
____

2. What is an acquired trait? Put the definition in your own words. ______________________
____
____
____

3. List four examples of inherited traits seen in animals. ______________________
____
____
____

4. List four examples of acquired traits seen in animals. ______________________
____
____
____

Directions: Complete the chart with your traits. List at least four inherited traits and four acquired traits.

Some of my inherited traits	Some of my acquired traits

Inherited & Acquired Traits Animal Coloring

Directions: Read each animal trait. Identify the trait as an inherited or an acquired trait by coloring the animal accordingly. Then answer the question at the bottom of the page using complete sentences.

1

Trait: An owl has excellent night vision.

If it's an inherited trait, color the owl orange. If it's an acquired trait, color the owl red.

2

Trait: A dog quickly runs to a person who yells, "Treat!"

If it's an inherited trait, color the dog green. If it's an acquired trait, color the dog purple.

3

Trait: A mouse develops a strong preference for cheese.

If it's an inherited trait, color the mouse pink. If it's an acquired trait, color the mouse brown.

4

Trait: An octopus can open a jar to get what is inside.

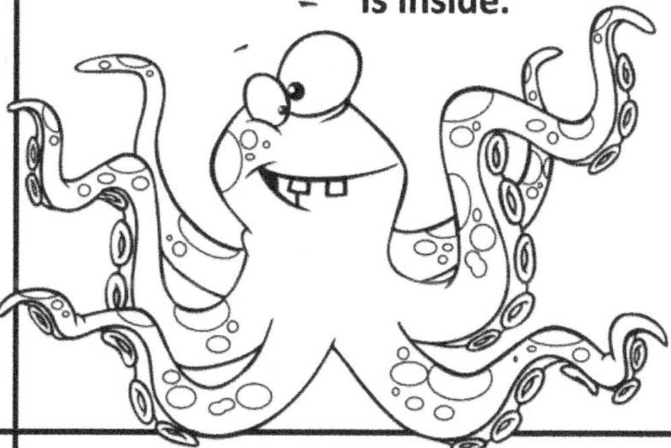

If it's an inherited trait, color the octopus blue. If it's an acquired trait, color the octopus red.

5

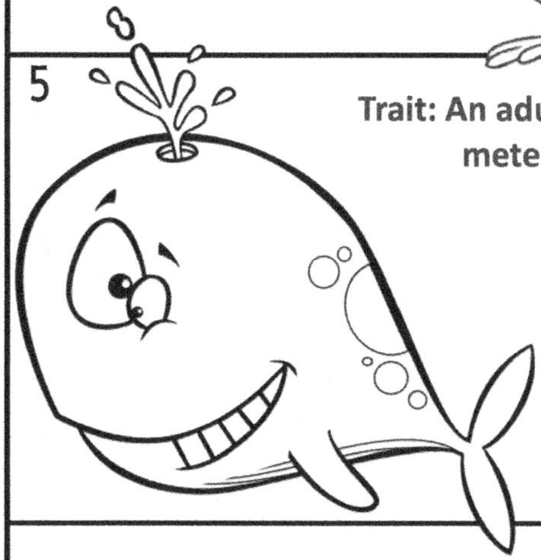

Trait: An adult whale is 30 meters long.

If it's an inherited trait, color the whale blue. If it's an acquired trait, color the whale orange.

6

Trait: A squirrel limps after breaking its leg as a youngster.

If it's an inherited trait, color the squirrel green. If it's an acquired trait, color the squirrel yellow.

What is the difference between inherited and acquired traits? ______________________

____

____

____

____

____

____

Dog Traits
Card Sorting Activity

Directions:

1. Cut out the cards on page 99.
2. Sort the cards into the chart below.
3. Check your answers.
4. Explain your answers to another person. You can use this sentence frame to guide you.
   → "I can tell the ____________________ card goes in the __________________ column of the chart because __________________."

Inherited Traits	Acquired Traits

Dog Traits Card Sorting Activity

Directions:

1. Carefully cut out the cards below.
2. Sort the cards into the correct columns of the chart on page 98.
3. Follow the rest of the directions on page 98.

A corgi has a docked (shortened) tail.	A dog runs to the food bowl when she hears kibble being scooped.
An Australian shepherd has one blue eye and one brown eye.	A dog only poops and pees outside.
A dog sits when commanded to sit.	A Dalmatian is born deaf in both ears.
A dachshund has wiry hair.	Boston terriers and French bulldogs have short snouts.
An untrained border collie puppy starts to herd people.	A dog whines and wags his tail when it is time for a walk.

My Own Inherited Traits

Directions: Answer the questions below about your inherited traits by circling yes or no. If you have a slight mix, choose the answer most similar to your trait.

1. Do you have a cleft chin? *(Your chin might have a little dip or dimple in the middle. This is a cleft chin. If there is no indentation in the middle, you have a smooth chin.)*

   Yes No

2. Do you have at least one dimple?

   Yes No

3. Do you have Darwin's tubercule? *(Look at the upper half of the outer rim of your ear. Is there a little bump that sticks out? If so, that's a Darwin's tubercule.)*

   Yes No

4. Are your earlobes detached? *(Detached earlobes hang free. Attached earlobes are connected to the side of the head.)*

   Yes No

5. Do you have freckles?

   Yes No

6. Do you have curly hair?

   Yes No

7. Do you have a widow's peak hairline? *(You can have a smooth hairline or a widow's peak, which is a V-shaped hairline.)*

   Yes No

8. Do you have straight thumbs? *(If you give a thumbs up, is your thumb straight or does it bend backward? A curved thumb is called a hitchhiker's thumb.)*

   Yes No

9. Can you roll your tongue?

   Yes No

10. Are you right-handed?

    Yes No

Make a Prediction:
What percentage of your ten traits are dominant?

My Own Inherited Traits

Directions: Use your answers from page 101 to determine how many of your traits are dominant and how many are recessive. Every *yes* answer is a dominant trait and every *no* answer is a recessive trait.* Place a checkmark in the appropriate column for each trait. Then answer the rest of the questions.

My Inherited Traits		
Trait	Dominant	Recessive
Trait 1: Cleft chin		
Trait 2: Dimples		
Trait 3: Darwin's tubercule		
Trait 4: Earlobes		
Trait 5: Freckles		
Trait 6: Hair curl		
Trait 7: Hairline		
Trait 8: Thumbs		
Trait 9: Tongue rolling		
Trait 10: Handedness		

What percent of your traits are dominant?

What percent of your traits are recessive?

Do your results surprise you? Why or why not? Answer in sentences. ______________________________

____

____

____

____

____

____

____

____

____

**There is still much to learn about human traits. Some of these inherited traits may be polygenic or involve codominance or incomplete dominance or other factors. Therefore, it is possible for a mix of traits to appear. If a mix appears, choose the answer that most closely resembles your trait.*

Inherited Traits Survey Activity

Directions: Use this list of questions to ask others about their inherited traits. Include a minimum of ten people, more if possible. Record everyone's answers with tally marks in the appropriate columns in the chart.

1. Do you have a cleft chin?
   Yes (dominant) No (recessive)
2. Do you have at least one dimple?
   Yes (dominant) No (recessive)
3. Do you have Darwin's tubercule?
   Yes (dominant) No (recessive)
4. Are your earlobes detached?
   Yes (dominant) No (recessive)
5. Do you have freckles?
   Yes (dominant) No (recessive)
6. Do you have curly hair?
   Yes (dominant) No (recessive)
7. Do you have a widow's peak hairline?
   Yes (dominant) No (recessive)
8. Do you have straight thumbs?
   Yes (dominant) No (recessive)
9. Can you roll your tongue?
   Yes (dominant) No (recessive)
10. Are you right-handed?
    Yes (dominant) No (recessive)

How many people did you include in your survey?

Frequencies of Inherited Traits

Trait	Dominant (yes)	Recessive (no)
1: Cleft chin		
2: Dimples		
3. Darwin's tubercule		
4. Earlobes		
5. Freckles		
6. Hair curl		
7. Hairline		
8. Thumbs		
9. Tongue rolling		
10. Handed-ness		

Inherited Traits Survey Activity

Directions: Create a bar graph to show the results of your survey. On the y-axis, add your numbered intervals. (Are you counting by 1s, 2s, 5s, or 10s? If you surveyed a small number of people, it makes sense to have an interval of one or two. If you surveyed more than thirty people, it might make better sense to have an interval of five or ten.) On the x-axis, *D* stands for dominant and *R* stands for recessive. The trait number and its description are underneath each *D* and *R*.

Frequencies of Dominant & Recessive Inherited Traits

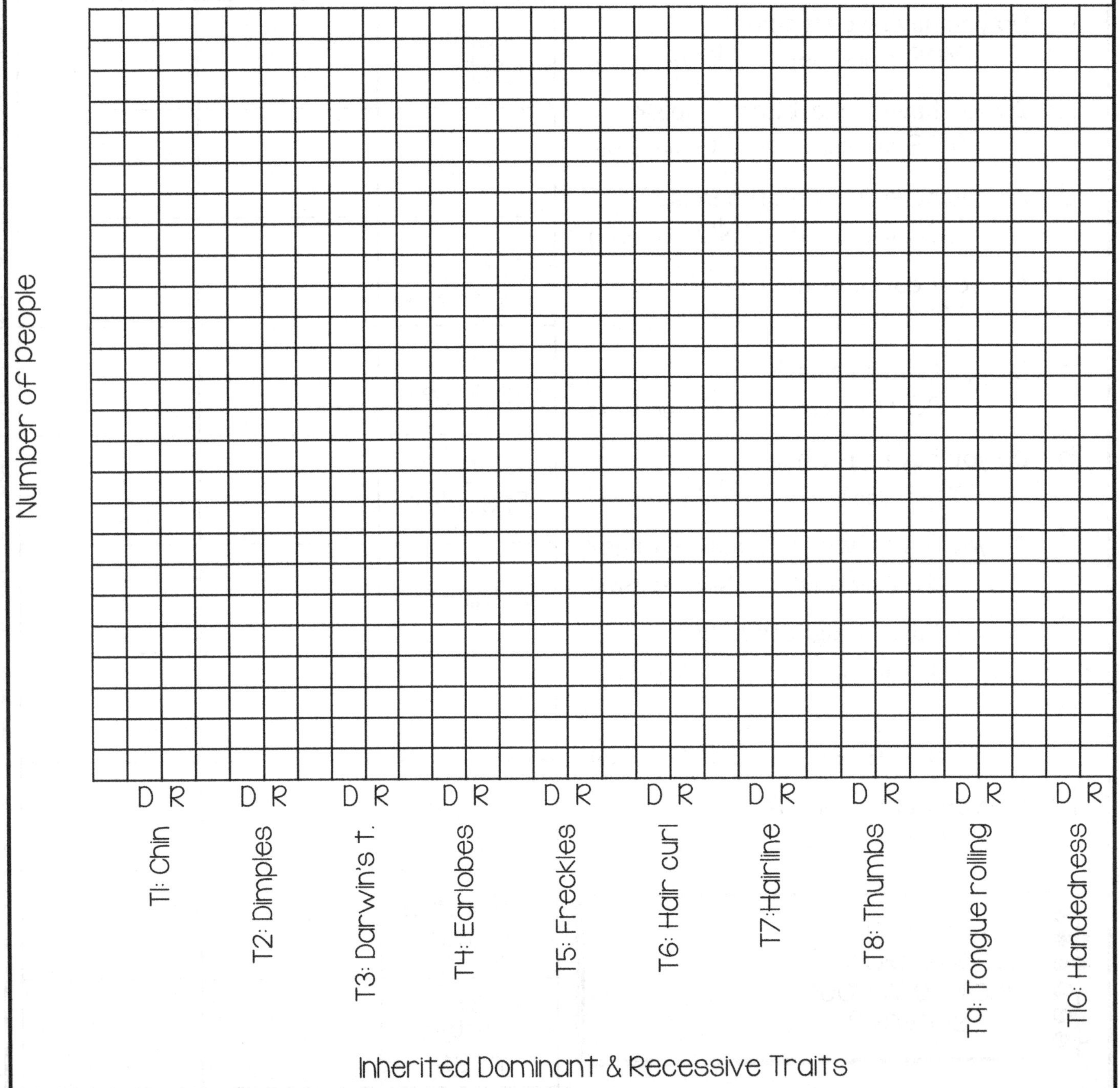

Inherited Traits Survey Activity Data Analysis

Directions: Use the results from your graph of inherited traits on page 104 to complete this page.

1. Briefly write about the people you surveyed. How do you know them? Are they from diverse or similar backgrounds? ____________________

2. Examine your graph. Write about three trends or details you see in the data. ____________________

3. Compare the results of your own trait survey on page 102 to the results you graphed. What similarities and differences do you see? ____________________

4. Based on this small survey, do dominant traits seem more common than recessive traits? Use examples from your data to support your answer. ____________________

5. If you surveyed more people and were able to include people from all over the world, from all sorts of backgrounds, how do you think your survey results would change? Explain your answer.

Research a Trait

Choose a human trait to research. Use the guiding questions to write about this trait in paragraph form.

Guiding Questions:

- How common is this trait?
- Is this trait inherited, acquired, or both?
- If the trait is inherited, is it known whether it is dominant or recessive?
- If the trait is acquired, how is it acquired?
- What do scientists know about this trait?
- What is still unknown about this trait?
- Have there been any recent discoveries about this trait?

Trait Choices

- Having red hair
- Having albinism
- Sneezing in bright light (ACHOO syndrome)
- Being extroverted
- Having an anxiety disorder
- Other:

Inherited & Acquired Traits in People

1. Where do inherited traits come from? Answer in a sentence. ____________________
____
____
____

2. Where do acquired traits come from? Answer in a sentence. ____________________
____
____
____

3. Look at the traits listed below. Underline the inherited traits. Circle the acquired traits.

Becoming afraid of spiders after being bit	Having the ability to roll your tongue
Loving Italian food	Thinking pink is a color only for girls
Having naturally brown hair	Throwing a football with a tight spiral
Having a family history of webbed toes	Being a cat person
Having dimples	Your blood type

4. List two inherited traits and two acquired traits that can help someone become a successful actor or actress.

Inherited trait: ______________________________

Inherited trait: ______________________________

Acquired trait: ______________________________

Acquired trait: ______________________________

5. Think about one of your friends or family members. List two inherited traits that make them special. Then list two of your favorite acquired traits they have.

Inherited trait: ______________________________

Inherited trait: ______________________________

Acquired trait: ______________________________

Acquired trait: ______________________________

Inherited & Acquired Traits Assessment

1. True or False: People are born with all of the traits they will ever have.

2. What is the best definition of an inherited trait?
   a. It is a trait received from experiences.
   b. It is a trait received because of a genetic mutation.
   c. It is a trait received genetically from parents.
   d. It is a trait that can only be received from formal education.

3. What is the best definition of an acquired trait?
   a. It is a trait received from experiences.
   b. It is a trait received because of a genetic mutation.
   c. It is a trait received genetically from parents.
   d. It is a trait that can only be received from formal education.

4. You notice a little girl with celery green eyes. What kind of trait is this? How do you know? Explain your answer in sentences. ______________________________

5. A bodybuilder has huge muscles. What kind of trait is this? How do you know? Explain your answer in sentences. ______________________________

Codominance & Incomplete Dominance

Predict the meaning of codominance.

- Take some time to think about what codominance might mean and write a possible definition of this term. Come up with a genetic example you can see in your life that might demonstrate codominance.

Predict the meaning of incomplete dominance.

- Take some time to think about what incomplete dominance might mean and write a possible definition of this term. Come up with a genetic example you can see in your life that might demonstrate incomplete dominance.

Codominance & Incomplete Dominance

When Things Aren't Mendelian

Gregor Mendel introduced us to dominant and recessive alleles and the fact that only the dominant trait is seen in heterozygous offspring (when two different types of alleles are paired). But, adding complexity to genetics, not all heterozygous allele pairs behave this way. Sometimes the two opposing alleles are equally dominant to each other. Other times, the two opposing alleles are neither dominant nor recessive. What happens in those cases?!

Exploring Codominance with Toddlers

Picture a living room with two unruly toddlers, one with a red crayon and one with a blue crayon. They are unsupervised and there is a nice white wall to color. Grab a red and blue crayon and, in the sketch pad below, draw what the wall will look like when these toddlers have finished making their "art."

You probably drew a wall with harsh scribbles of red interspersed with bold streaks of blue. The toddler with the red represents one dominant allele. The toddler with the blue represents a different dominant allele. The wall with both red and blue represents the phenotype of a codominant trait. The red and the blue did not mix together; both colors were very, very visible to the parents when they made their way to the living room three minutes later.

In codominance, <u>both</u> alleles are dominant so <u>both</u> traits are seen in heterozygous offspring. Think about a dominant trait—it demands to be seen if the allele is there and is not going to take no for an answer. Codominance happens in heterozygous pairs when both of the alleles are dominant, making both traits visible in the phenotype.

Codominance & Incomplete Dominance

Painting Incomplete Dominance

Picture two gentle, loving children. One child has red paint. The other child has blue paint. There is only one canvas.

The child with red paint turns to the other and says, “It’s okay. You can have the canvas.”

The child with the blue paint responds, “No, it’s alright. *You* can have it. Please. I insist.”

“I have an idea. We’ll blend our red and blue paint, and we will paint the canvas purple together!” exclaims the child with the red paint.

In this example, the child with red represents one allele and the child with blue represents another allele. The purple canvas represents the phenotype of a trait showing incomplete dominance.

In incomplete dominance, the alleles are neither dominant nor recessive, so the traits blend together in heterozygous offspring. To remember the meaning of incomplete dominance, remember the following phrase:

<u>In</u>complete dominance goes <u>in</u> the bender.

When ingredients get put in a blender, they mix up and blend together. You don’t see the different ingredients anymore—it’s just a nice blend of what was tossed in there. Same with incomplete dominance.

Punnett Square Showing Mendelian Inheritance

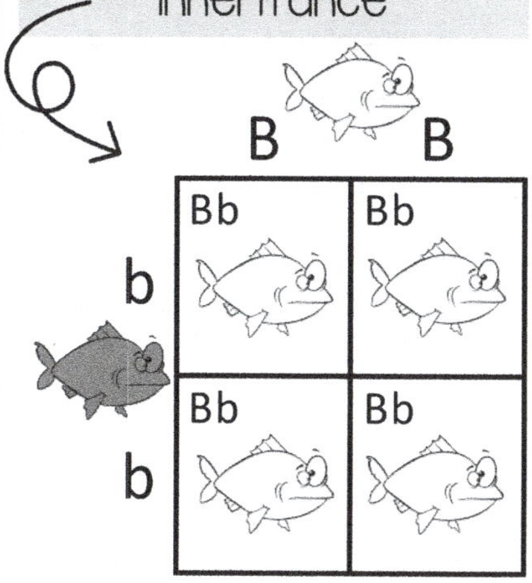

Punnett Square Showing Codominance

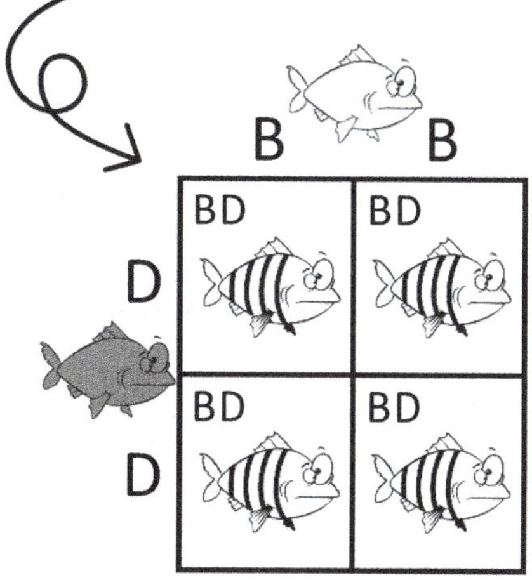

Punnett Square Showing Incomplete Dominance

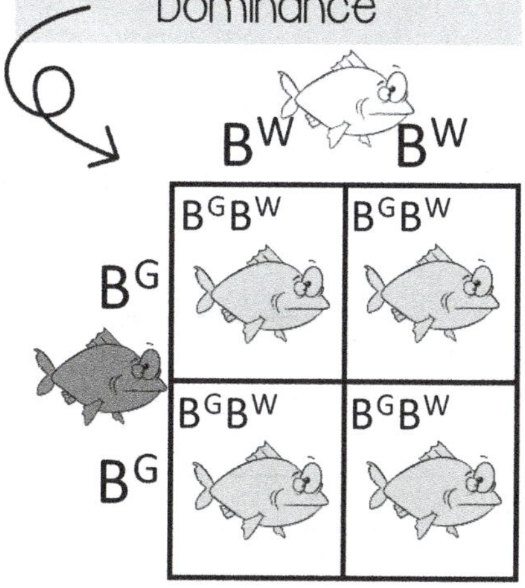

Codominance & Incomplete Dominance Practice

Directions: Answer the following questions in complete sentences.

1. What is codominance? ______

2. What is incomplete dominance? ______

3. Can codominance and incomplete dominance be seen in homozygous allele pairs or heterozygous allele pairs? ______

Directions: Identify the examples below as codominance or incomplete dominance. Then explain your answer in a sentence.

4. Yellow flower x Red flower = Orange flower ______

5. White flower x Red flower = White flower with red spots ______

6. Straight hair x Curly hair = Wavy hair ______

7. Red cow x White bull = Roan calf (roan means both colors are seen in the hair) ______

8. Long tail x Short tail = Medium length tail ______

Coloring Heterozygous Unicorns

Directions: Color the unicorns in a way that shows Mendelian inheritance, codominance, or incomplete dominance. Use the provided genetic information to determine your color choices.

Mendelian Inheritance:
Twinkle McSparkleface has a genotype of Rr for his coloration. One of his parents is rainbow (RR) and the other is pink (rr).

Codominance:
Neighoncé has a genotype of YB for her coloration. One of her parents is yellow (YY) and the other is blue (BB).

Incomplete Dominance:
Bubbles the Destroyer has a genotype of C^RC^W for his coloration. One of his parents is white (C^W) and the other is red (C^R).

Codominance & Incomplete Dominance Cat Coloring

Directions: Determine if the text is describing codominance or incomplete dominance. Circle your answer. Then use the two listed colors to color the cat in a way that shows your circled answer. Finally, answer the question at the bottom of the page using complete sentences.

1

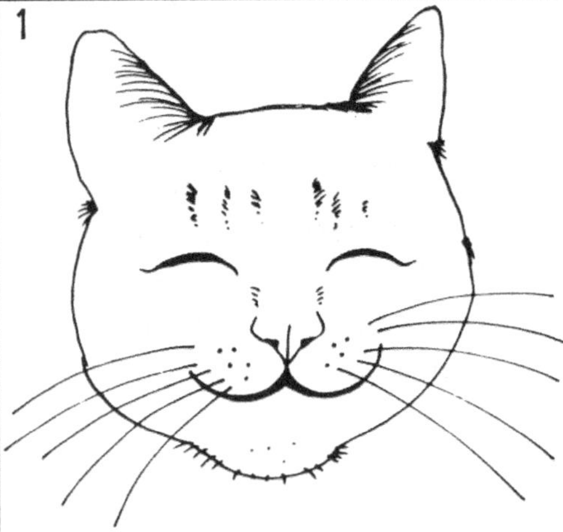

A heterozygous offspring has alleles that are neither dominant nor recessive so the traits mix together.

Codominance or Incomplete Dominance?

Red and Blue

2

A frog has black and yellow spots, showing the traits of two dominant alleles at the same time.

Codominance or Incomplete Dominance?

Black and White

3

Codominance or Incomplete Dominance?

Yellow and Red

A heterozygous offspring shows the traits of both dominant alleles in its phenotype.

4

Codominance or Incomplete Dominance?

Yellow and Blue

A flower is a nice pink color, blending the traits it received from a red parent and a white parent.

What is the difference between codominance and incomplete dominance? ____________

____

____

____

____

____

____

Codominance & Incomplete Dominance Assessment

1. What is codominance?
    a. It is when a dominant allele and a recessive allele are paired, but only the dominant trait is seen in the offspring.
    b. It is when both alleles are dominant, so both traits are seen in heterozygous offspring.
    c. It is when the alleles are neither dominant nor recessive, so they blend together into a phenotype that is different from either parent.
    d. It is when a homozygous allele pair expresses its dominant trait.

2. What is incomplete dominance?
    a. It is when a dominant allele and a recessive allele are paired, but only the dominant trait is seen in the offspring.
    b. It is when both alleles are dominant, so both traits are seen in heterozygous offspring.
    c. It is when the alleles are neither dominant nor recessive, so they blend together into a phenotype that is different from either parent.
    d. It is when a homozygous allele pair expresses its dominant trait.

Directions: Classify the examples of codominance and incomplete dominance by circling your answer.

3. A dark purple eggplant and a white eggplant create a light violet eggplant.

    Codominance　　　　Incomplete dominance

4. A rabbit with long hair and a rabbit with short hair create a rabbit with medium-length hair.

    Codominance　　　　Incomplete dominance

5. A white chicken and a black chicken create a black-and-white checkered chicken.

    Codominance　　　　Incomplete dominance

6. A yellow tulip and a red tulip create a tulip with swirls of yellow and red.

    Codominance　　　　Incomplete dominance

Genetics Vocabulary Review Card Game

Overview

This card game is used to practice and strengthen students' science vocabulary. It follows similar rules and gameplay as Taboo™. It can be played in a variety of ways. Detailed instructions are on the next page. The cards are found on pages 119-130.

Number of Players

This game can be played with two or more people.

Ideas for Use

These cards can be used in many ways. Here are some ideas for how to use them.

- Students can work in pairs or small groups to play with the cards in the deck together.
- Use these cards as a quick review at the beginning or end of a lesson. The learning partner or teacher can provide the clues to the student(s) as the card-reader, or the student(s) can be the card-reader.
- Go through the whole deck before the genetics assessment as a fun review.
- The learning partner or teacher can state the vocabulary word on the top of a card and instruct the student(s) to come up with a list of taboo words for that vocabulary word. Compare the card with the students' list to see how many match.
- Students can make their own cards using the blank cards found on page 129.

Genetics Vocabulary Review Card Game

How to Play

This game follows similar gameplay as Taboo™. Start by cutting out all of the cards on pages 119-130. Then you will be ready to play using the following objective and rule.

Objective

Card-readers need to get the other players to guess the word or phrase in the shape at the top of the card.

Rule

The card-reader cannot say any form of the words below the shape at the top of the card.

Add Your Own Rules

- Choose a time limit for correctly guessing the card.
- Decide whether or not the card-reader can skip a card.
- See how many cards players can get correct within a set amount of time.

Guess this word on top

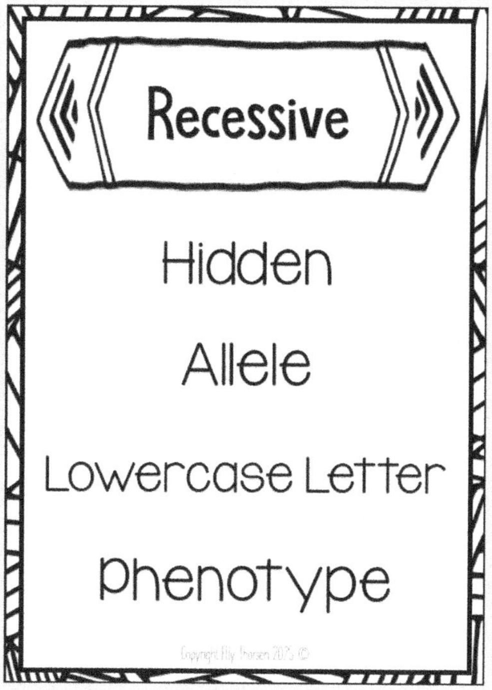

Don't say the words below

Examples of what card-readers CAN say:

- "This is the opposite of dominant."
- "It is not shown with capitals."
- "You only see its physical trait in a homozygous pair."

Examples of what card-readers CANNOT say:

- "It is **hidden** whenever it is paired with a dominant **allele**."
- "It is represented with **lowercase letters**."
- "This kind of **allele** only shows its **phenotype** in a homozygous pair."

Genetics

Heredity

Study

Traits

Genes

Heredity

Generation

Parents

Offspring

Pass

Trait

Characteristic

Parents

Offspring

Physical

Codominance

Both

Dominant

Heterozygous

Trait

Incomplete Dominance

Blend

Trait

Heterozygous

Mix

Purebred

Same

Generation

Homozygous

Trait

Hybrid

Different

Dissimilar

Heterozygous

Type

Gregor Mendel

Genetics

Pea Plants

Monk

Father

Gene

Allele

Trait

Inherit

Parent

Allele

Gene

Dominant

Recessive

Form

Dominant

Seen

Allele

Phenotype

Capital Letter

Recessive

Hidden

Allele

Lowercase Letter

Phenotype

Punnett Square

Predict

Box

Offspring

Percent

Genotype

Allele

Combination

Genes

Letter

Phenotype

Physical

Trait

Appearance

Organism

Homozygous

Same

Allele

Identical

Purebred

Heterozygous

Different

Allele

Hybrid

Pair

Acquired Trait

Learned

Taught

Experience

Injury

Inherited Trait

Heredity

Genes

Characteristic

Parent

Offspring

Baby

Parent

Generation

Young

Genetics Vocabulary & Concepts Review

Part One Directions: Use the word bank to fill in the blanks. Each word will be used once.

Word Bank		
Genotype	Phenotype	Homozygous
Heterozygous	Heredity	Hybrid
Purebred	Dominant	Recessive
Genetics	Incomplete dominance	Punnett square

1. The scientific study of heredity is called ______________________.
2. ______________________ is the passing of characteristics from parents to offspring.
3. Whenever a ______________________ allele is present its trait will show up.
4. Geneticists often use a ______________________ to calculate the frequencies of different genotypes and phenotypes in offspring.
5. A ______________________ is the physical appearance or the traits of an organism.
6. A ______________________ is the offspring of two distinctly different parental types.
7. When an organism has two of the same alleles for a trait it is considered ______________________.
8. A ______________________ allele can only show up if there are no dominant alleles present.
9. In ______________________ the alleles are neither dominant nor recessive so a mix of both traits is seen in the offspring.
10. The allele combination or genetic makeup of an organism is called its ______________________.
11. If an organism is ______________________ it has two different alleles for a trait.
12. An offspring with many generations of the same trait is a ______________________.

Genetics Vocabulary & Concepts Review

Part Two Directions: Complete each question below using your knowledge and understanding of genetics.

13. Circle the example(s) of genotype.

    HH   Brown eyes   Curly hair   gg

14. Provide an example of a phenotype. ______________________________

15. Circle the example(s) of a heterozygous allele pair.

    Rr   LL   mm

16. Provide an example of a homozygous allele pair. ______________________________

17. Circle the example(s) that would show a dominant trait.

    GG   Ww   bb

18. Provide an example of a recessive genotype. ______________________________

19. If both a dominant and a recessive allele are present, which trait will show up and why? Answer in sentences. ______________________________

20. A gardener crossed a white flower with a red flower. He expected some offspring to be white and others to be red. Instead, all of the offspring were white with red spots. What occurred and how can you tell? Answer in sentences. ______________________________

Genetics Odd One Out

Directions: Analyze each part of the circle. One of the parts will be different than the others. Shade the part of the circle that does not match. Using a complete sentence, explain how that part is different from the others. *Answers must be science related and should show your understanding of genetics and heredity.*

Example

Chemistry	**History**
Biology	Physics

History is the odd one out because it is not a type of science like the others.

1

RR	ff
Dominant	Hh

2

Old English Sheepdog	Golden retriever
Mixed-breed dog	German shepherd

3

A white fish with black spots	A white fish
A black fish	A green fish

4

aa	TT
ee	Qq

5

Having attached earlobes	Being likely to develop allergies
Having naturally curly hair	Speaking the English language

6

Homozygous	Heterozygous
Hybrid	Yy

Fairy Tale Genetics

Directions: Solve the genetic questions using a Punnett square. Use capital and lowercase letter B for your Punnett squares. Then complete each table with the possible genotypes, phenotypes, and the percent chance of each offspring type. The first one is completed for you.

1. In unicorns, swirly horns are dominant to plain horns. If a heterozygous unicorn is crossed with a purebred dominant unicorn, what are the possible genotypes and phenotypes of the offspring and the percent chance for each? Complete the Punnett square to help you find your answers.

	B	b
B	BB	Bb
B	BB	Bb

Genotype	Phenotype	Percent Chance
BB	Swirly horns	2 out of 4 = 50% chance
Bb	Swirly horns	2 out of 4 = 50% chance
bb	Plain horns	0 out of 4 = 0% chance

2. In gnomes, hairy feet are dominant to hairless feet. If a homozygous dominant gnome is crossed with a homozygous recessive gnome, what are the possible genotypes and phenotypes of the offspring and the percent chance for each? Complete the Punnett square to help you find your answers.

Genotype	Phenotype	Percent Chance

3. In dragons, fire-breathing is dominant to ice-breathing. If an ice-breathing dragon is crossed with a hybrid fire-breathing dragon, what are the possible genotypes and phenotypes of the offspring and the percent chance for each? Complete the Punnett square to help you find your answers.

Genotype	Phenotype	Percent Chance

Fairy Tale Genetics

4. In werewolves, dark brown hair is a dominant trait and silver hair is a recessive trait. If a silver haired werewolf is crossed with a purebred werewolf with dark brown hair, what are the possible genotypes and phenotypes of the offspring and the percent chance for each? Complete the Punnett square to help you find your answers.

Genotype	Phenotype	Percent Chance

5. In mermaids, emerald-colored tails are dominant to rainbow-colored tails. Two mermaids with heterozygous genotypes are crossed together. What are the possible genotypes and phenotypes of the offspring and the percent chance for each? Complete the Punnett square to help you find your answers.

Genotype	Phenotype	Percent Chance

6. Create your own crossing using fairies, elves, centaurs, zombies, vampires, or other creature of your choosing. Write an example like the ones from above and solve it using the Punnett square. Fill in the genotype, phenotype, and percent chance table. Then, in the box, draw an offspring with the dominant phenotype showing.

Genotype	Phenotype	Percent Chance

Genetics Stations Activity

Activity Options:

There are multiple options for how to complete this activity.

- Option 1: Cut out the station cards on pages 137-148 and post the cards around a given area. There are two choices for this option.
  - The cards can be hidden by the teacher or learning partner and the student can search for them. Once the student finds the cards, the answers can be written directly on the cards or the student can explain the answers aloud to another person.
  - The cards can be taped in numerical order around a room, up and down a hallway, or outside. The student should visit each card and write the answers directly on the card or explain the answers aloud to another person.
- Option 2: Cut out the station cards on pages 137-148 and stack them in a pile. The student and the teacher or learning partner can use the cards together like flashcards. Alternatively, the student can complete the cards by themselves by writing the answers directly on the cards.
- Option 3: Do not cut out the cards. Instead, complete the cards exactly how they are in this workbook.

Extension Idea:

The student can make their own station cards using the blank cards found on page 149.

What is a recessive allele?

1.

Examine the genotypes. Which are homozygous? How do you know?

RR Gg Ll mm HH Yy

2.

Directions: Cut out each of the cards.

How likely is it that a Bb father and a bb mother will have bb offspring?

3.

Which trait will show up if both a dominant and a recessive allele are present?

4.

A female bird is homozygous for large eyes, which is a dominant trait. Her mate is homozygous for small eyes. What percent of their offspring are likely to have small eyes?

5.

A liger is a cross between a tiger and a lion. A mule is the offspring of a male donkey and a female horse. A puggle is half pug and half beagle. What are ligers, mules, and puggles considered in genetics?

6.

Directions: Cut out each of the cards.

If being tall is a dominant trait and being short is a recessive trait, will a plant with a genotype of Tt be tall or short? Explain your answer in a sentence.

7.

What is a genotype?

8.

Give three examples of inherited traits in people.

9.

Give three examples of acquired traits in people.

10.

Directions: Cut out each of the cards.

Shellton John is a turtle heterozygous for his medium-green shell color. His mom is a purebred with a dark green shell. His dad is a purebred with a pale green shell. Is Shellton an example of Mendelian inheritance, codominance, or incomplete dominance? How do you know?

11.

What is codominance?

12.

Do capital or lowercase letters represent dominant alleles?

13.

A male cat has recessive blue eyes, and his genotype is ee. What genotype(s) could a female cat have in order to produce kittens that share the father's phenotype?

14.

Directions: Cut out each of the cards.

Your friend has a guinea pig with long hair named Piggy Smalls. Piggy's parents, grandparents, and great-grandparents all had long hair. Is Piggy a purebred or a hybrid? Explain your answer in sentences.

15.

Give three examples of phenotypes.

16.

Tulips can come in solid colors like yellow and red. Sometimes, however, the colors in tulips are swirled together and two colors are seen at the same time. What is this an example of? Explain in sentences.

17.

Which part of the Punnett square contains a mistake?

part 1 — part 2

	R	r
R	RR	RR
r	Rr	rr

part 3 — part 4

18.

Directions: Cut out each of the cards.

What are alleles?

19.

Who is considered the father of modern genetics?

20.

How are purebreds and hybrids different?

21.

Use the chart to answer the question. Two parents both have a heterozygous genotype for hair color. How likely are their children to have blonde hair?

Dominant Trait	Recessive Trait
Dark hair	Blonde hair
Curly hair	Straight hair
Widow's peak hairline	Straight hairline

22.

Directions: Cut out each of the cards.

Use the chart to answer the question. A woman has naturally straight hair. A man has curly hair but is not homozygous for that trait. How likely is their baby to inherit curly hair?

Dominant Trait	Recessive Trait
Dark hair	Blonde hair
Curly hair	Straight hair
Widow's peak hairline	Straight hairline

23.

Use the chart to answer the question. One parent has two dominant alleles for hairline. The other parent has a dominant and a recessive allele for hairline. How likely are their children to have a widow's peak?

Dominant Trait	Recessive Trait
Dark hair	Blonde hair
Curly hair	Straight hair
Widow's peak hairline	Straight hairline

24.

Make your own genetics and heredity card.

Make your own genetics and heredity card.

Directions: Cut out each of the cards.

Make your own genetics and heredity card.

Make your own genetics and heredity card.

Genetics Test

Part I Directions: Use the word list to complete the sentences. You will not use all of the words. None of the words will be used more than once.

Word List				
Alleles	Codominance	Recessive	Genetics	Genes
Genotype	Heredity	Hybrid	Traits	Heterozygous
Purebred	Dominant	Phenotype	Homozygous	Incomplete dominance

1. A ______________________________ is an offspring with many generations of the same trait, like a beagle or a pug.
2. The different forms of the genes are called ______________________________.
3. The allele whose trait always shows up in the offspring if it is present is called ______________________________. It is represented with capital letters.
4. A ______________________________ is the allele combination or genetic makeup of an organism. An example is RR.
5. The allele that is hidden whenever a dominant allele is present is called ____________________.
6. A ______________________________ is an organism's physical appearance. An example is brown hair.
7. In ______________________________ the alleles are neither dominant nor recessive. An example of this is when a black rodent is crossed with a white rodent and their offspring are gray.
8. The scientific study of heredity is known as ______________________________.
9. An organism with two different alleles for a trait, such as Rr, is ______________________________.
10. An organism with two of the same alleles for a trait, such as RR, is ______________________________.

Genetics Test

Part 2 Directions: Read the statements and circle True or False.

11. **True** or **False**: A recessive trait will not be seen if a dominant and recessive allele are paired together.

12. **True** or **False**: An organism with a genotype of Rr is most likely a purebred.

13. **True** or **False**: GG and Gg are both examples of a homozygous genotype.

14. **True** or **False**: Gregor Mendel is known as the father of genetics.

15. **True** or **False**: Having green eyes is an example of a phenotype.

Part 3 Directions: Circle the letter of your answer.

16. What happens to the recessive allele when a dominant allele is present?
    a. The recessive allele mixes with the dominant allele and both traits are seen.
    b. The recessive allele disappears completely and only the dominant allele is seen.
    c. The recessive allele takes over and covers the dominant allele.
    d. The recessive allele is there but it is hiding and only the dominant allele's trait is seen.

17. What is the difference between genotype and phenotype?
    a. Genotype is the genetic makeup of an organism and is represented with letters, while phenotype is the trait of an organism and is often visible.
    b. Phenotype is the genetic makeup of an organism and is represented with letters, while genotype is the trait of an organism and is often visible.
    c. Genotype and phenotype mean the same thing.
    d. Genotype is when an organism has two of the same alleles for a trait, and phenotype is when an organism has two different alleles for a trait.

18. In a plant with a genotype of Tt (with T being dominant and t being recessive), what trait or traits will show up?
    a. Only the dominant trait will appear.
    b. Only the recessive trait will appear.
    c. Both the dominant and recessive traits will be seen.
    d. The dominant and recessive traits will blend together.

Genetics Test

19. Examine the genetic cross shown in the Punnett square. In domesticated rabbits, black hair is dominant (B) and white hair is recessive (b). What hair color are the offspring of this crossing expected to have?

    a. 100% of the offspring will have black hair.
    b. 100% of the offspring will have white hair.
    c. 75% of the offspring will have black hair and 25% will have white hair.
    d. 25% of the offspring will have black hair and 75% will have white hair.

	B	b
B	BB	Bb
b	Bb	bb

20. Which of the examples below shows codominance?

    a. A small Chihuahua crossed with a large spaniel makes a medium-sized dog.
    b. A curly haired mother and a straight-haired father have a child with wavy hair.
    c. An orange cat is crossed with a white cat and their kittens are white with orange markings.
    d. All of the above are examples of codominance.

21. Which of the examples below is a trait a person can inherit genetically?

    a. Having long fingers
    b. Being able to play a certain lullaby on a piano
    c. Having a crooked finger after an accident with a car door
    d. Having a tan line on your finger from wearing a ring all summer

22. Which of the examples below is an acquired trait?

    a. Having short eyelashes
    b. Having an eyelid that droops after an injury
    c. Having one blue eye and one brown eye
    d. Being nearsighted and having trouble seeing

23. The Punnett square contains a mistake. Where is the mistake located?

    a. The mistake is in quarter 1.
    b. The mistake is in quarter 2.
    c. The mistake is in quarter 3.
    d. The mistake is in quarter 4.

	H	h
H	HH (1)	HH (2)
H	HH (3)	Hh (4)

Genetics Test

Part 4 Directions: Read the genetic problems. For each problem, complete the Punnett square and write a sentence to answer the question.

24. An Ee parent is crossed with an ee parent. What percent of their offspring will have an ee genotype?

25. A BB parent is crossed with a Bb parent. B (dominant) is for brown eyes and b (recessive) is for blue eyes. What percent of their offspring will have brown eyes?

26. A rabbit is a dominant purebred for large ears and it is crossed with a rabbit homozygous for small ears (a recessive trait). What percent of their offspring will have small ears?

27. You know your pet mouse has a recessive homozygous allele pair because it has red eyes. Your friend's mouse has black eyes, and every mouse it is related to has black eyes as well. If your mouse and your friend's mouse were crossed, what percent of their offspring will have black eyes?

Genetics Test

28. Having freckles is a dominant trait. Two parents heterozygous for this trait are expecting a baby. What percent chance will the baby have of carrying the recessive allele and having the potential to pass it on to future generations?

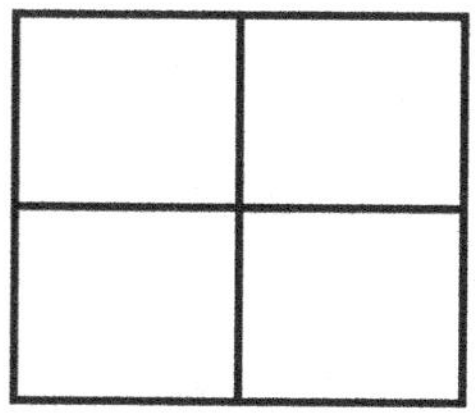

29. In people, Huntington's disease is dominant. Rashid does not have dominant alleles for this trait and will never develop Huntington's disease. Leilani does have one dominant allele for this trait and will develop Huntington's disease. If Rashid and Leilani have a baby, what are the chances their baby will get two recessive alleles and therefore not develop Huntington's disease?

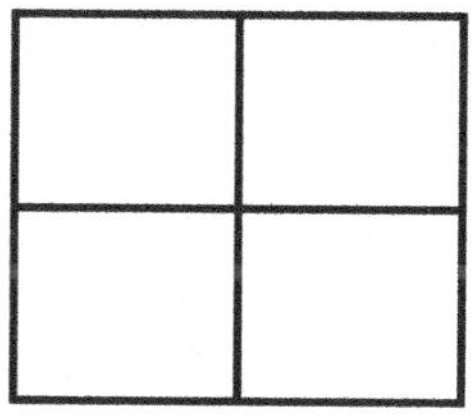

Part 5 Directions: Read the problem and respond thoroughly with a paragraph.

30. Mariana bred a female Labrador Retriever with a male standard poodle. Based on what she knows about dominant and recessive alleles, she predicted some puppies would have the smooth hair of their mother and some puppies would have the curly hair of their father. However, the puppies had wavy hair that was a mix of both of their parents' hair types. Use what you know about genetics to explain to Mariana why she got this unexpected result.

Ideas to Explore Next

Learning More:

Do you want to learn more about genetics? Here are some topics to check out next.

- Mendel's Principle of Segregation
- Mendel's Principle of Independent Assortment
- Genetic mutations
- Real life hybrid animals like pizzly bears, beefalos, ligers, and Savannah cats
- Genetically Modified Organisms (GMOs)
- DNA
- Chromosomes
- Sex-linked chromosomes and sex-linked disorders like hemophilia and red-green color blindness
- Gene mapping
- Gene therapy and its benefits and controversies
- Cloning
- Selective breeding in pets and farm animals
- Darwin's theory of evolution and how it relates to Mendelian genetics
- Walter Sutton (Here's an analogy for you: Gregor Mendel was to peas as Walter Sutton was to grasshoppers.)
- Famous geneticists from history and the present day
- Epigenetics
- Population genetics
- Eugenics in history and present day—for older or more mature students
- New genetic discoveries that happened this year

Hands-On Activities & Research Ideas

Dive Deeper:

Here are some activities and research ideas to continue learning about genetics and heredity.

- Learn about and create bigger Punnett squares—dihybrid and trihybrid crosses instead of the monohybrid crosses that appear in this workbook.
- Find your blood type and the blood type of your family with a blood type testing kit.
- Get a DNA genetic testing kit to learn more about your ancestry.
- Play a genetics game such as *Genotype* or *Variation*.
- Research a dog or cat breed. What genetic diseases or issues are they likely to inherit? Why?
- Research a genetic disorder such as cystic fibrosis, sickle cell anemia, or Huntington's disease.
- Make a family tree.
- Watch a movie or documentary about genetics.
- Pick a visible trait such as hairline, earlobes, or chin type. Take pictures of all of your willing family members and track that trait throughout your family.
- Find examples of codominance and incomplete dominance in plants around your geographic area or in plants at a local nursery or garden center.
- Do a deep dive into guppy genetics and see which varieties you can spot at a local pet store.
- Interview a dog, cat, rat, or livestock breeder to learn about how genetics shapes their breeding program.
- Research and form your own opinions on the ethical issues surrounding genetic testing.
- Examine your fingerprints and compare them to your family members, noting any similarities.
- Make a comic or write a story about genetics.

Glossary of Terms

Acquired Trait
An acquired trait is also called a learned trait. It is a trait that is not passed from parents to offspring genetically. Instead, acquired traits are from experiences and/or environmental factors.

Alleles
Alleles are the different forms of the genes. Alleles might be dominant or recessive.

Codominance
In codominance both of the alleles are dominant, so both traits are seen in the phenotype of heterozygous offspring. Example: A yellow flower is crossed with a red flower and their offspring are yellow with red stripes. You see *both* the yellow and the red.

Dominant
A dominant allele is one whose trait always shows up if the allele is present. It is represented with capital letters.

Genetics
Genetics is the scientific study of heredity.

Genes
Genes are the factors that control the traits or characteristics of an organism. Genes are the basic units of heredity passed from parents to offspring. They contain information about traits.

Genotype
A genotype is the allele combination or genetic makeup of an organism. Examples of genotypes include rr, heterozygous, and homozygous dominant.

Heredity
Heredity is the passing of traits from parents to offspring.

Heterozygous
Heterozygous means two different alleles for a trait. Sometimes, this means having one dominant allele paired with one recessive allele. Heterozygous might be written as Rr.

Homozygous
Homozygous means having two of the same alleles for a trait. This can mean having two dominant alleles paired together or having two recessive alleles paired together. Homozygous might be written as RR or rr.

Hybrid
A hybrid is the offspring of two distinctly different parental types. It is an organism with two different alleles for a trait (heterozygous). Hybrid might be written as Rr.

Incomplete Dominance
In incomplete dominance the alleles are neither dominant nor recessive, so they mix together in the phenotype of heterozygous offspring. Example: A yellow flower is crossed with a red flower and their offspring are orange. You see a *blend* of the yellow and the red.

Glossary of Terms

Inherited Trait
An inherited trait is a trait that is passed genetically from parents to offspring.

Mendelian Inheritance
Mendelian inheritance (or simple inheritance) is defined as genetically getting a trait controlled by a single gene that comes in dominant or recessive. Traits inherited according to Mendel's principles are called Mendelian.

Monogenic
Monogenic means a trait is controlled by only one gene.

Offspring
Offspring are the new organisms produced by a living thing. Offspring are the babies.

Organism
An organism is any individual living thing.

Parent
An organism that contributes genetic material to offspring.

Phenotype
Phenotype is an organism's observable characteristics or the traits it has. Examples of phenotypes include hair color, susceptibility to disease, and height.

Polygenic
Polygenic means a trait is controlled by multiple genes working together.

Principle of Dominance
This principle states some alleles are dominant and others are recessive; therefore, in a heterozygous genotype, the trait of one allele can hide the trait of the other allele. In other words, when an organism has two different alleles, the dominant trait will be expressed and the recessive trait will be masked.

Punnett Square
A Punnett square is a diagram used to predict the genetic makeup of offspring.

Purebred
A purebred is an organism with many generations of the same trait. It has two of the same alleles for a trait (homozygous). Purebred might be written as RR or rr.

Recessive
A recessive allele is one whose trait will only show up if there is no dominant allele around. It is represented with lowercase letters.

Traits
Traits are the characteristics or features passed from parents to offspring.

Answer Key

An Introduction to Genetics (Page 13)

1. Offspring are the new organisms produced by living things. Examples may include seedlings and puppies.
2. Traits are the characteristics passed from parents to offspring. Examples may include eye color and height.
3. Heredity is the passing of characteristics from one generation to the next.
4. Genetics is the study of heredity.
5. Gregor Mendel was an Austrian monk who conducted extensive experiments on pea plants. His discoveries form the foundation of our current understanding of genetics.
6. The theory of blending inheritance describes traits mixing together in offspring.
7. According to blending inheritance, a tall parent and a short parent would have offspring of medium height because the heights mixed together.

Mendel's Pea Plants (Page 15)

Predictions will vary.

Mendel's Flower Color Experiment (Page 18)

- 100% of the offspring had purple flowers. The white flowers seemed to disappear. *(The flower in the center of the page should be colored purple.)*
- 75% of the offspring were purple and 25% of the offspring were white. *(The bottom row of four flowers should have the first three colored purple and the last flower white.)*

Mendel's Pea Plants: A Summary (Page 19)

Summary paragraphs will vary. Example paragraph:

At the beginning of Mendel's experiment, he crossed a purebred tall pea plant with a purebred short pea plant. This P generation produced the F_1 generation of hybrid pea plants, which were all tall. When the F_1 generation was crossed with each other, they produced the F_2 generation. The F_2 generation had some surprising results. Seventy-five percent of the F_2 generation were tall. Twenty-five percent were short; this short trait that wasn't present in the F_1 generation returned.

Dominant & Recessive Practice (Page 22)

1. Genes are basic units of heredity passed from parents to offspring. They contain information about traits and are located within cells. *(If you want to be really specific, you can say genes are located on chromosomes which are found in the nucleus of a cell.)*
2. Alleles are the different forms of a gene. In a pair of alleles, one allele comes from each parent.
3. Alleles can be dominant or recessive. Dominant alleles have the ability to mask the traits of recessive alleles.
4. A dominant allele is one whose trait always shows up if it is there.
5. A recessive allele is one whose trait can only be seen if no dominant allele is present. A recessive allele's trait is hidden if a dominant allele is around.
6. The dominant trait
7. The recessive trait
8. The dominant trait

Answer Key

Celebrity Symbols of Dominant & Recessive Alleles (Page 23)

1. Dominant alleles
2. Two dominant alleles
3. One dominant allele and one recessive allele
4. The dominant trait is seen.
5. A recessive trait can only be seen if two recessive alleles are paired together.

- Dominant celebrity example: *A big, beefy actor can represent a dominant allele because he is strong and takes over whenever he's in a scene, just like a dominant allele takes over when it is present in a gene.*
- Recessive celebrity example: *A quiet child actor can represent a recessive allele because he is in the background and is hardly noticeable unless no grown-up actors are present, just like a recessive allele's trait is only seen when no dominant alleles are present in a gene.*

Dominant & Recessive Alleles Flower Coloring (Page 24)

1. Yellow—dominant
2. Red—recessive
3. Red—recessive
4. Yellow—dominant
5. Yellow—dominant
6. Red—recessive

- Answers may vary. Example answer: *If a dominant allele is around it takes over everything. If a recessive allele is around it won't show up unless it is paired with another recessive allele.*

Dominant & Recessive Card Sorting Activity (Pages 25-28)

The dominant trait will be seen.

- Two dominant alleles paired together
- One dominant allele and one recessive allele paired together
- Nn
- GG
- Two capital letters
- A capital and a lowercase letter
- A hybrid
- The offspring has two purebred parents with opposite traits

The recessive trait will be seen.

- Two recessive alleles paired together
- rr
- Two lowercase letters

Trick answer: The offspring has two purebred parents with the same trait. *This could be in either column depending on if both parents are purebred dominant or if both parents are purebred recessive.*

Answer Key

Dominant & Recessive Extra Practice (Page 29)

1. Capital
2. Lowercase
3. Two dominant alleles paired together *and* One dominant allele paired with one recessive allele
4. Two recessive alleles paired together
5. Uu, Ll, AA, BB
6. rr, ii, nn
7. Is there a recessive allele? → *Yes! It's the y in Yy.*
   Is the recessive trait seen? → *No! A dominant allele is present so the recessive allele hides.*
   Is the recessive allele gone completely **or** is it lurking in the background while the dominant allele gets noticed? → *The recessive allele is lurking unseen in the background!*

Dominant Might Not Equal Common (Page 34)

The sentences students highlight on page 33 will vary.

1. Example answer: *Dominant traits are not necessarily more common than recessive traits. If there are not many dominant alleles in a population to begin with, those dominant alleles might remain uncommon unless they are beneficial in some way.*
2. Example answer: *One reason why a dominant trait might be less common than a recessive trait is if that trait makes survival more difficult. An example is if a dominant trait makes it hard to absorb nutrients from food. The inability to get necessary nutrients makes it harder for that organism to thrive and pass on the dominant trait to future generations, allowing the recessive trait to become common and stay common.*
3. Example answer: *When a trait helps an organism survive and/or reproduce, it becomes more common over time. This is because the organisms who have that trait are more likely to live long enough to reproduce and give that trait to future generations, gradually making that trait more common.*
4. Example answer: *When a trait makes it more difficult to survive and/or reproduce, that trait is most likely to become less common over time. This is because the organisms who have that trait are less likely to live long enough to reproduce and give that trait to future generations, gradually making that trait less common.*
5. Example answer: *When a trait is in an undisturbed population and does not impact survival and reproduction, that trait is likely to remain stable over time. It has no reason to increase or decrease, so its prevalence is unlikely to have much fluctuation.*
6. Example answer: *Two examples of genetic drift are the bottleneck effect and the founder effect. These both typically involve small populations.*

Dominant & Recessive Assessment (Page 35)

1. True
2. Dominant
3. Recessive
4. D → Some students may pick B, but that is incorrect because the recessive allele is still in the organism's genes. The recessive allele does not disappear—its trait just is not seen in the organism.
5. A
6. True
7. False

Answer Key

Genotype & Phenotype (Page 36)

1. TT, Tt, and tt
2. Tall and short
3. TT or Tt
4. tt
5. All of the tall pea plants have the same phenotype, but they can have a different genotype.
6. All short pea plants have the same genotype and the same phenotype. *Genotype and phenotype can be in either order in this answer.*

Genotype & Phenotype Practice (Page 37)

1. Example answer: *Genotype is the genetic makeup of an organism. Examples of genotype include Yy and two recessive alleles.*
2. Example answer: *Phenotypes are the characteristics or traits of an organism. Examples of phenotype include dark green scales and small leaves.*
3. True. This statement is true because both fish have the same triangle pattern on their dorsal fins, which is an example of phenotype.
4. False. This statement is false because there are two possible genotypes for the fish to have the dominant triangle pattern. The fish could have two dominant alleles for this trait or they could have a dominant and a recessive allele for this trait. Both of those genotypes would display the triangle pattern.

Genotype & Phenotype Worm Coloring (Page 38)

1. Green—genotype
2. Purple—phenotype
3. Purple—phenotype
4. Green—genotype
5. Purple—phenotype
6. Green—genotype

- Two organisms can have the same phenotype but different genotypes. If organisms have the dominant phenotype, they can have a genotype of two dominant alleles *or* a genotype of one dominant allele and one recessive allele.

Genotypes Plus Adorable Phenotypes (Page 39)

- The box on the left should show a guinea pig with short hair. This guinea pig's genotype is GG or Gg. This guinea pig's phenotype is short hair.
- The box on the right should show a guinea pig with long hair. This guinea pig's genotype is gg. This guinea pig's phenotype is long hair.
- Example answer: *The difference between genotype and phenotype is that genotype involves the genes and phenotype involves the expressed traits. In the guinea pig example, the genotypes are GG, Gg, and gg. The phenotypes, or visible characteristics, are long hair and short hair. Genotype relates to genes. Phenotype relates to traits that are often visible.*

Answer Key

Genotype & Phenotype Odd One Out (Page 40)

****Answers will vary. It is possible for a question to have more than one correct answer. The following are example correct answers.*

1. Visible traits. This is the odd one out because it is about phenotype. All of the others are about genotype.
2. Heart-shaped leaf. This is the odd one out because it is a phenotype. All of the others are genotypes.
3. Physical characteristics. This is the odd one out because it is about phenotype. All of the others are about genotype.
4. Tall. This is the odd one out because it is about phenotype. All of the others are about genotype.
5. Hybrid. This is the odd one out because it is a genotype. All of the others are phenotypes.
6. A dog's ancestors are all the same breed. This is the odd one out because it is a genotype. All of the others are phenotypes.

Genotype & Phenotype Assessment (Page 41)

1. Genotype
2. Phenotype
3. A
4. Genotype
5. Phenotype
6. Phenotype
7. Genotype
8. Genotype
9. Genotype
10. Phenotype

Practice with Punnett Squares (Page 44)

2. 0 out of 4 are AA, 0% chance of AA
   2 out of 4 are Aa, 50% chance of Aa
   2 out of 4 are aa, 50% chance of aa
3. 4 out of 4 are EE, 100% chance of EE
   0 out of 4 are Ee, 0% chance of Ee
   0 out of 4 are ee, 0% chance of ee
4. 0 out of 4 are MM, 0% chance of MM
   2 out of 4 are Mm, 50% chance of Mm
   2 out of 4 are mm, 50% chance of mm
5. 1 out of 4 are BB, 25% chance of BB
   2 out of 4 are Bb, 50% chance of Bb
   1 out of 4 are bb, 25% chance of bb
6. 2 out of 4 are FF, 50% chance of FF
   2 out of 4 are Ff, 50% chance of Ff
   0 out of 4 are ff, 0% chance of ff

2.

	A	a
a	Aa	aa
a	Aa	aa

3.

	E	E
E	EE	EE
E	EE	EE

4.

	m	m
M	Mm	Mm
m	mm	mm

5.

	B	b
B	BB	Bb
b	Bb	bb

6.

	F	F
F	FF	FF
f	Ff	Ff

Answer Key

Identifying Incorrect Punnett Squares (Page 45)

1. Incorrect
2. Incorrect
3. Correct
4. Incorrect
5. Correct
6. Incorrect

1.

	B	b
b	Bb	bb
b	~~BB~~	bb

Bb

2.

	L	L
l	Ll	Ll
l	Ll	~~ll~~

Ll

4.

	r	r
R	Rr	Rr
r	~~Rr~~	rr

rr

6.

	N	n
N	NN	Nn
n	~~NN~~	nn

Nn

Predicting Dominant Traits with Punnett Squares (Page 46)

2. There is a 100% chance of producing the dominant calm phenotype.
3. There is a 100% chance of producing the dominant green phenotype.
4. There is a 75% chance of producing the dominant smooth phenotype.

2.

	R	R
R	**RR**	**RR**
r	**Rr**	**Rr**

3.

	R	R
r	**Rr**	**Rr**
r	**Rr**	**Rr**

4.

	R	r
R	**RR**	**Rr**
r	**Rr**	rr

Remember, any letter can be used to create the Punnett squares—just focus on the letter being capital or lowercase. These examples were completed with letter R. The genotypes that will show the dominant phenotype are in bold.

Homozygous & Heterozygous Practice (Page 64)

1. Homozygous means having two of the same alleles for a trait.
2. Heterozygous means having two different alleles for a trait.
3. A pair of dominant alleles is homozygous because two of the same alleles are together.
4. A genotype of one dominant allele and one recessive allele is heterozygous because two different alleles are together.
5. AA, uu, two dominant alleles, two recessive alleles
6. Bb, hybrid
7. In order to show a recessive trait, a genotype will be homozygous. The only way a recessive trait can be seen is if two recessive alleles are together. Since two of the same alleles are paired, this is homozygous.
8. In order to show a dominant trait, a genotype will be homozygous or heterozygous. A dominant trait can be seen if two dominant alleles are paired (homozygous) or if a dominant and a recessive allele are paired (heterozygous).

Homozygous & Heterozygous Corgi Coloring (Page 65)

1. Blue—heterozygous
2. Orange—homozygous
3. Blue—heterozygous
4. Orange—homozygous
5. Orange—homozygous
6. Blue—heterozygous

- Heterozygous genotypes have two different alleles paired together. Homozygous genotypes have two of the same alleles paired together.

Answer Key

Homozygous & Heterozygous Punnett Squares (Pages 66-68)

1. HH correctly represents a homozygous dominant genotype because it has two of the same dominant alleles paired together.
2. Hh correctly represents a heterozygous genotype because it has two different alleles paired together, one dominant allele and one recessive allele.
3. hh correctly represents a homozygous recessive genotype because it has two of the same recessive alleles paired together.
4. *Already solved on pages 66 and 68*
5. 100% chance of homozygous dominant. 0% chance of heterozygous. 0% chance of homozygous recessive. 100% of the offspring are expected to show the dominant trait.
6. 0% chance of homozygous dominant. 0% chance of heterozygous. 100% chance of homozygous recessive. 0% of the offspring are expected to show the dominant trait.
7. 25% chance of homozygous dominant. 50% chance of heterozygous. 25% chance of homozygous recessive. 75% of the offspring are expected to show the dominant trait.
8. 0% chance of homozygous dominant. 50% chance of heterozygous. 50% chance of homozygous recessive. 50% of the offspring are expected to show the dominant trait.
9. 50% chance of homozygous dominant. 50% chance of heterozygous. 0% chance of homozygous recessive. 100% of the offspring are expected to show the dominant trait.
10. 0% chance of homozygous dominant. 100% chance of heterozygous. 0% chance of homozygous recessive. 100% of the offspring are expected to show the dominant trait.
11. 0% chance of homozygous dominant. 50% chance of heterozygous. 50% chance of homozygous recessive. 50% of the offspring are expected to show the dominant trait.

5.

	E	E
E	EE	EE
E	EE	EE

6.

	e	e
e	ee	ee
e	ee	ee

7.

	E	e
E	EE	Ee
e	Ee	ee

8.

	e	e
E	Ee	Ee
e	ee	ee

9.

	E	e
E	EE	Ee
E	EE	Ee

10.

	e	e
E	Ee	Ee
E	Ee	Ee

11.

	E	e
e	Ee	ee
e	Ee	ee

Colorful Monster Phenotypes with Provided Genotypes (Page 69)

1. hh genotype and yellow star phenotype
2. RR genotype and red tail phenotype
3. Ll genotype and orange and green spots phenotype
4. Jj genotype and green hair phenotype

1.

	h	h
H	Hh	Hh
h	hh	hh

2.

	R	R
R	RR	RR
r	Rr	Rr

3.

	L	L
l	Ll	Ll
l	Ll	Ll

4.

	J	j
j	Jj	jj
j	Jj	jj

Answer Key

Colorful Monster Phenotypes with Genotype Clues (Page 70)

Any letter can be used for the Punnett squares and the genotypes. Focus on whether they are dominant (capital) or recessive (lowercase).

1. Ee genotype and pink tusks phenotype
2. Ee genotype and red snail eyes phenotype
3. EE genotype and brown eyebrows phenotype
4. Ee genotype and blue tongue phenotype

1.

	E	E
E	EE	EE
e	Ee	Ee

2.

	e	e
E	Ee	Ee
e	ee	ee

3.

	E	E
E	EE	EE
e	Ee	Ee

4.

	E	e
E	EE	Ee
e	Ee	ee

Make Your Own Colorful Monsters (Page 71)

Answers will vary.

Homozygous & Heterozygous Experiment (Pages 72-75)

Answers will vary.

Famous Monster Genetics (Pages 76-77)

	Genotype	Phenotype	Percent Chance	Punnett Square
2.	BB	White fur	0 out of 4 = 0% chance	(B, b) / b: Bb, bb / b: Bb, bb
	Bb	White fur	2 out of 4 = 50% chance	
	bb	Gray fur	2 out of 4 = 50% chance	
3.	BB	Wiry hair	0 out of 4 = 0% chance	(b, b) / B: Bb, Bb / b: bb, bb
	Bb	Wiry hair	2 out of 4 = 50% chance	
	bb	Smooth hair	2 out of 4 = 50% chance	
4.	BB	Very shy	1 out of 4 = 25% chance	(B, b) / B: BB, Bb / b: Bb, bb
	Bb	Very shy	2 out of 4 = 50% chance	
	bb	Outgoing	1 out of 4 = 25% chance	
5.	BB	Average size	0 out of 4 = 0% chance	(b, b) / B: Bb, Bb / B: Bb, Bb
	Bb	Average size	4 out of 4 = 100% chance	
	bb	Large size	0 out of 4 = 0% chance	
6.	BB	Large feet	0 out of 4 = 0% chance	(B, B) / b: Bb, Bb / b: Bb, Bb
	Bb	Large feet	4 out of 4 = 100% chance	
	bb	Tiny feet	0 out of 4 = 0% chance	

Answer Key

Homozygous & Heterozygous Assessment (Page 79)

1. Heterozygous
2. Homozygous
3. False
4. False
5. Circle—homozygous
6. Circle—homozygous
7. Underline—heterozygous
8. Underline—heterozygous
9. Purebreds are homozygous. In order for the same trait to show up again and again, uninterrupted within the generations, it needs to be homozygous. A homozygous genotype has to be crossed with an identical homozygous genotype in order for the offspring to be a purebred.

Review of Genetics Concepts (Page 80)

Alleles—*They are the different forms of the genes. They might be dominant or recessive.*
Dominant—*It is the one whose trait always shows up if the allele is present.*
Genes—*They are the basic units of heredity that contain information about the traits passed from parents to offspring.*
Genetics—*It is the scientific study of heredity.*
Heredity—*It is the passing of traits from parents to offspring.*
Heterozygous—*It is when two different alleles are paired together.*
Homozygous—*It is when two of the same alleles are paired together.*
Hybrid—*It is an offspring with two distinctly different parental types.*
Gregor Mendel—*He is known as the father of modern genetics. No, you may not call him the Mendelorian.*
Genotype—*It is the allele combination or genetic makeup of an organism.*
The Principle of Dominance—*When an organism has two different alleles, the dominant trait will be expressed and the recessive trait will be masked.*
Purebred—*It is an organism with many generations of the same trait.*
Recessive—*Its trait only shows up when no dominant alleles are present.*
Traits—*They are the characteristics or features of an organism.*
It's seen when two recessive alleles pair.—*It's the recessive phenotype!*
It's seen when one dominant and recessive allele pair.—*It's the dominant phenotype!*

Dominant & Recessive Dog Traits (Page 81)

1. Dropped ears
2. Wiry fur
3. Brown eyes
4. Homozygous
5. EE
6. Bb
7. Ll and LL
8. cc

Answer Key

9. 0% chance of straight fur
10. A. 50% chance
    B. 75% chance

9.

	C	C
c	Cc	Cc
c	Cc	Cc

10.

	L	l
L	LL	Ll
l	Ll	ll

Genetics Vocabulary Review Volume 1 (Page 82)

Answers will vary.

Genetics Vocabulary Review Volume 2 (Page 83)

Answers will vary.

Farm Animal Genetics (Pages 84-86)

1. dd
2. Brown eyes
3. Genotype: bb Phenotype: Red hair
4. Yes
5. Hh and HH
6. Colorful feathers
7. ll
8. DD
9. Genotype: Hh Phenotype: Smooth hair
10. Mm and MM
11. 100%
12. *See pie chart*
13. 50%
14. 50%
15. 50% *see pie chart*
16. 0%
17. *Answers will vary.*

11.

	a	a
a	aa	aa
a	aa	aa

12.

	M	m
M	MM	Mm
m	Mm	mm

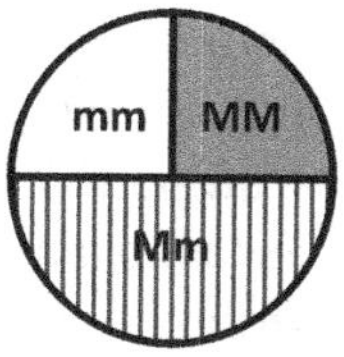

13.

	H	H
H	HH	HH
h	Hh	Hh

14.

	B	b
b	Bb	bb
b	Bb	bb

15.

	E	E
E	EE	EE
e	Ee	Ee

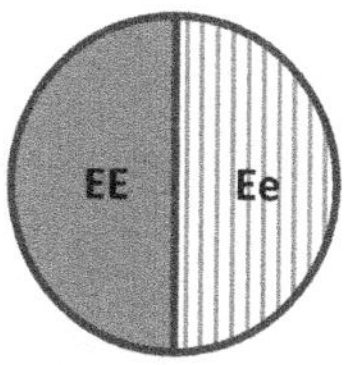

16.

	F	F
f	Ff	Ff
f	Ff	Ff

Make an Organism Project (Pages 87-91)

Answers will vary.

Beyond Just Dominant & Recessive Practice (Page 93)

1. Monogenic means a trait is controlled by just one gene or only two alleles.
2. Polygenic means a trait is controlled by multiple genes working together.
3. People are more complex than pea plants. Pea plant height is monogenic, meaning it's controlled by just one gene. The height of a person is determined by multiple genes, allowing for a much wider range of heights.
4. *Answers will vary* but might include hair texture, hair color, eye color, nose width, eyelash length, and skin tone. These seem likely to be polygenic because they are so varied among people.
5. A person genetically predisposed to allergies might never develop allergies if they're never exposed to the right environmental factors or specific allergens.

Answer Key

Getting Your Traits (Page 94)

Answers will vary.

Inherited & Acquired Traits (Page 95)

An Incredible Athlete's Traits:

- Inherited Traits: height and shoulder width
- Acquired Traits: technique and size/strength of muscles
- Mixed Traits: sense of competition and drive to win

Inherited & Acquired Traits Practice (Page 96)

1. An inherited trait is one that is passed genetically from parents to offspring.
2. An acquired trait, also called a learned trait, is one that is gained through experiences and environmental factors.
3. *Answers will vary* but may include large ears, growing 42 teeth, having stripes, and having scales.
4. *Answers will vary* but may include having a chipped tooth, walking with a limp after an injury, avoiding a certain plant after getting sick from eating it, and using a stick as a tool.

- *Answers will vary.*

Inherited & Acquired Traits Animal Coloring (Page 97)

1. Inherited trait—orange
2. Acquired trait—purple
3. Acquired trait—brown
4. Acquired trait—red
5. Inherited trait—blue
6. Acquired—yellow

- Inherited traits are received genetically from parents to offspring. Acquired traits are received from experiences and situations.

Dog Traits Card Sorting Activity (Pages 98-99)

Inherited Traits

- An Australian shepherd has one blue eye and one brown eye.
- A Dalmatian is born deaf in both ears.
- A dachshund has wiry hair.
- Boston terriers and French bulldogs have short snouts.
- An untrained border collie puppy starts to herd people.

Acquired Traits

- A corgi has a docked (shortened) tail.
- A dog runs to the food bowl when she hears kibble being scooped.
- A dog only poops and pees outside.
- A dog sits when commanded to sit.
- A dog whines and wags his tail when it is time for a walk.

Answer Key

My Own Inherited Traits (Pages 101-102)

Answers will vary.

Inherited Traits Survey Activity (Pages 103-105)

Answers will vary.

Inherited & Acquired Traits in People (Page 107)

1. Inherited traits come from the genes received from parents.
2. Acquired traits come from experiences and the environment.
3. Inherited traits: Having the ability to roll your tongue, Having naturally brown hair, Having a family history of webbed toes, Having dimples, Your blood type
   Acquired traits: Becoming afraid of spiders after being bit, Loving Italian food, Thinking pink is a color only for girls, Throwing a football with a tight spiral, Being a cat person
4. *Answers will vary.* Example inherited traits include having an expressive face and having a specific type of natural beauty. Example acquired traits include being able to speak with different accents and being in good shape from exercise.
5. *Answers will vary.* Example inherited traits include having bright blue eyes and dimples when they smile. Example acquired traits include being good at writing funny stories and being a great goalie.

Inherited & Acquired Traits Assessment (Page 108)

1. False
2. C
3. A
4. Having celery green eyes is an inherited trait. Eye color is a trait received genetically from parents, so it is considered an inherited trait.
5. Having huge muscles is an acquired trait. Increasing muscle size is achieved through experiences such as weightlifting, so it is considered an acquired trait.

Codominance & Incomplete Dominance Practice (Page 112)

1. Codominance means both of the alleles are dominant, so both traits are seen in the phenotype of heterozygous offspring.
2. Incomplete dominance means the alleles are neither dominant nor recessive, so they mix together in the phenotype of heterozygous offspring.
3. Codominance and incomplete dominance can be seen in heterozygous allele pairs.
4. This is an example of incomplete dominance because the yellow and red *mixed* to make orange.
5. This is an example of codominance because the white and the red can *both* be seen.
6. This is an example of incomplete dominance because the two hair types *mixed* together to create wavy.
7. This is an example of codominance because *both* the red and white hair are seen in the offspring.
8. This is an example of incomplete dominance because the lengths *mixed* together to make a medium length.

Answer Key

Coloring Heterozygous Unicorns (Page 113)

- Twinkle McSparkleface should be rainbow colored.
- Neighoncé should have some yellow parts and some blue parts.
- Bubbles the Destroyer should be pink.

Codominance & Incomplete Dominance Cat Coloring (Page 114)

1. Incomplete dominance—purple cat
2. Codominance—black and white cat
3. Codominance—yellow and red cat
4. Incomplete dominance—green cat

- In codominance, both alleles are dominant so both traits are visible in the offspring. In incomplete dominance, the alleles are neither dominant nor recessive so the traits mix together in the offspring.

Codominance & Incomplete Dominance Assessment (Page 115)

1. B
2. C
3. Incomplete dominance
4. Incomplete dominance
5. Codominance
6. Codominance

Genetics Vocabulary & Concepts Review (Pages 131-132)

1. Genetics
2. Heredity
3. Dominant
4. Punnett square
5. Phenotype
6. Hybrid
7. Homozygous
8. Recessive
9. Incomplete dominance
10. Genotype
11. Heterozygous
12. Purebred
13. HH and gg
14. *Answers will vary.* Example answers include brown eyes and curly hair.
15. Rr
16. *Answers will vary.* Example answers include BB and two recessive alleles.
17. GG and Ww
18. Answers will vary. Example answer include hh and ee
19. If both a dominant and a recessive allele are paired, the dominant trait will show up because a dominant allele is present and takes over.

Answer Key

20. Codominance occurred with these flowers. Both the white and red were dominant, so both of those traits were seen in the resulting offspring. And, to dive deeper into this example, the gardener started with purebred white and purebred red flowers, which is why *all* of the flower offspring were white and red and not just some of the offspring.

Genetics Odd One Out (Page 133)

****Answers will vary. It is possible for a question to have more than one correct answer. The following are example correct answers.*

1. ff. This is the odd one out because it is recessive. All of the others have to do with the dominant allele or will show the dominant trait.
2. Mixed breed dog. This is the odd one out because it is not a purebred.
3. A white fish with black spots. This is the odd one out because it shows codominance with both traits visible. The others just show one trait.
4. Qq. This is the odd one out because it is heterozygous while all the others are homozygous.
5. Speaking the English language. This is the odd one out because it is an acquired trait and not an inherited trait like the other three.
6. Homozygous. This is the odd one out because it refers to two of the same alleles for a trait. The others refer to two different alleles for a trait.

Fairy Tale Genetics (Pages 134-135)

	Genotype	Phenotype	Percent Chance	Punnett Square
2.	BB	Hairy feet	0 out of 4 = 0% chance	B, B
	Bb	Hairy feet	4 out of 4 = 100% chance	b: Bb, Bb
	bb	Hairless feet	0 out of 4 = 0% chance	b: Bb, Bb
3.	BB	Fire-breathing	0 out of 4 = 0% chance	b, b
	Bb	Fire-breathing	2 out of 4 = 50% chance	B: Bb, Bb
	bb	Ice-breathing	2 out of 4 = 50% chance	b: bb, bb
4.	BB	Dark brown hair	0 out of 4 = 0% chance	b, b
	Bb	Dark brown hair	4 out of 4 = 100% chance	B: Bb, Bb
	bb	Silver hair	0 out of 4 = 0% chance	B: Bb, Bb
5.	BB	Emerald-colored tail	1 out of 4 = 25% chance	B, b
	Bb	Emerald-colored tail	2 out of 4 = 50% chance	B: BB, Bb
	bb	Rainbow-colored tail	1 out of 4 = 25% chance	b: Bb, bb

6. *Answers will vary.*

Answer Key

Genetics Stations Activity (Pages 136-148)

1. A recessive allele is one whose trait will only show up if there is no dominant allele around.
2. RR, mm, and HH are homozygous. I know because they either both have dominant or both have recessive—not dominant and recessive paired together.
3. 50% chance
4. The dominant trait
5. 0% of the offspring
6. Hybrids
7. A plant with a genotype of Tt will be tall because a dominant allele is present and masks the recessive trait.
8. A genotype is the allele combination or genetic makeup of an organism.
9. *Answers will vary* but may include dimples, brown eyes, and curly hair.
10. *Answers will vary* but may include having a chipped tooth, speaking two languages, and writing in cursive.
11. Shellton John's shell color is an example of incomplete dominance because the traits of his parents blended together to give him his medium-green shell color.
12. Codominance means both of the alleles are dominant, so both traits are seen in the phenotype of heterozygous offspring.
13. Capital letters
14. Ee or ee
15. Piggy Smalls is a purebred, which is an offspring with many generations of the same trait. Since all of Piggy's ancestors had long hair, he is considered a purebred.
16. *Answers will vary* but may include tall, black hair, and smooth seeds.
17. This is an example of codominance because both of the colors are seen in the offspring.
18. Part 2
19. Alleles are the different forms of the genes.
20. Gregor Mendel
21. Purebreds have many generations of the same trait and are homozygous. Hybrids are the result of unalike parents being crossed and are heterozygous.
22. 25% chance of having blonde hair
23. 50% chance of having curly hair
24. 100% chance of having a widow's peak

Genetics Test (Pages 151-155)

1. Purebred
2. Alleles
3. Dominant
4. Genotype
5. Recessive
6. Phenotype
7. Incomplete dominance
8. Genetics
9. Heterozygous
10. Homozygous

Answer Key

11. True
12. False
13. False
14. True
15. True
16. D
17. A
18. A
19. C
20. C
21. A
22. B
23. B
24. 50% of their offspring. See Punnett square below.
25. 100% of their offspring. See Punnett square below.
26. 0% of their offspring. See Punnett square below. Any letter may be used.
27. 100% of their offspring. See Punnett square below. Any letter may be used.
28. 75% chance of carrying a recessive allele and being able to pass it on to future generations. See Punnett square below. Any letter may be used.
29. 50% chance of inheriting two recessive alleles and therefore not developing Huntington's disease. See Punnett square below. Any letter may be used.
30. When Mariana bred the lab and the poodle together, their puppies' hair was wavy. This is incomplete dominance. Because the hair traits were neither dominant nor recessive, they blended together into hair unlike either parent.

24.

	E	e
e	Ee	ee
e	Ee	ee

25.

	B	B
B	BB	BB
b	Bb	Bb

26.

	R	R
r	Rr	Rr
r	Rr	Rr

27.

	r	r
R	Rr	Rr
R	Rr	Rr

28.

	F	f
F	FF	Ff
f	Ff	ff

29.

	h	h
H	Hh	Hh
h	hh	hh

About the Author:

Elly Thorsen attended the University of South Dakota where she earned her BS in Elementary Education and MA in Technology for Education and Training. She has taught in South Dakota, Minnesota, Oklahoma, and South Korea. She has experience teaching many grade levels and subject areas in a variety of school settings but loves seventh grade science best.

Find all of my workbooks on Amazon!

https://www.amazon.com/author/ellythorsen

Find more educational resources for your student!

https://www.teacherspayteachers.com/store/elly-thorsen

https://ellythorsenteaching.blogspot.com/

Made in the USA
Monee, IL
13 September 2026

62976635R10098